Lunar Outpost

The Challenges of Establishing a Human Settlement on the Moon

Erik Seedhouse

Lunar Outpost

The Challenges of Establishing a Human Settlement on the Moon

 Springer

Published in association with
Praxis Publishing
Chichester, UK

Dr Erik Seedhouse, F.B.I.S., As.M.A.
Milton
Ontario
Canada

SPRINGER–PRAXIS BOOKS IN SPACE EXPLORATION
SUBJECT *ADVISORY EDITOR*: John Mason, M.Sc., B.Sc., Ph.D.

ISBN 978-0-387-09746-6 Springer Berlin Heidelberg New York

Springer is part of Springer-Science + Business Media (springer.com)

Library of Congress Control Number: 2008934751

Cover design: Jim Wilkie
Project management: Originator Publishing Services, Gt Yarmouth, Norfolk, UK

Printed on acid-free paper

Contents

Preface

My intention in writing this book was to write a narrative of the key mission architecture elements comprising NASA's plan for returning astronauts to the Moon. Although this book is by no means an exhaustive account of the many steps required to enable a manned lunar mission, my goal was to present as much detail as possible. To that end I relied extensively on a vast number of documents ranging from Power-point presentations, briefings, plans, press conferences, and technical articles.

Only a generation ago, the United States abandoned its own pioneering space exploration program. Even as the Apollo 17 astronauts returned from the Moon in 1972, the Nixon Administration was closing the hatch on missions beyond low Earth orbit. The Saturn V fleet, together with all the technological wonders developed by NASA to fly astronauts to the Moon, were mothballed. Since the end of Apollo, thousands of scientific papers and popular articles have been written on the topic of returning humans to the Moon. But how will we actually return? The book you are now holding answers that question and is written for those who wonder how NASA's new fleet of Launch Vehicles are developing, what the new class of astronauts will do on the surface of the Moon, and how mission profiles are designed.

As I write these words, a Presidential election is looming and it would be remiss not to address the historic choices faced by the two candidates. Whether driven by political tactics or strategic statesmanship, the space policy decisions made by either Senator Barack Obama or Senator John McCain may well determine the direction the United States takes in its second half-century of manned spaceflight. Whereas under the Bush Administration manned spaceflight has been ascendant, comments made by Obama, such as those below, suggest job losses in the human spaceflight business to be almost a certainty.

"I think it is important for us to inspire through the space program, but also have the practical sense of what investments deliver the most scientific and

technological spinoffs, and not just to feel that the human space exploration—
actually sending bodies into space—is always the best investment."

Senator Barack Obama speaking during a call-in interview with
the Editorial Board of the *Houston Chronicle*

In contrast to Obama's position on space, McCain's space policy favors a return to the
Moon and preparations for manned missions to Mars, a policy reflected on his website
which carries an artist's rendering of the Orion Crew Exploration Vehicle. For seven
years, between 1997 and 2005, McCain was Chairman of the Senate Commerce
Committee, which oversees space and commercial aviation, a position that provided
him with experience critical to shaping NASA's future. As a strong supporter of
NASA and the space program, McCain is proud to have sponsored legislation
authorizing funding consistent with the President's VSE and he believes support
for a continued American presence in space is crucial to re-establishing the country's
pre-eminence over the Russians in space technology.

While some may worry about the potential ramifications of a possible Democratic
Administration for manned spaceflight, the reality is that no matter who wins the
White House in November 2008, both John McCain and Barack Obama will be long
gone, or well into a second term, before any policy changes can be implemented that
could seriously affect the Constellation Program. With fuel prices spiraling out of
control, an economy in recession, and a commitment to overhaul the education
system, the newly elected President will need to focus his attention on more pressing
concerns.

The Constellation Program, which is the focus of much of this book, is more than
just a proposal. It represents nothing less than the next logical phase in the evolution
of space exploration and the first step to creating a space-faring civilization. The
Vision for Space Exploration, announced by President Bush in January 2004, is a
vision that plans to send astronauts not only to the Moon but also onward to Mars
and beyond. It is a vision whose scope is vast. Far from being speculative, the plan
described in this book is being realized as you read these words.

Although the Constellation Program has its critics, NASA detractors would be
wise to remember that the agency invented the American manned space program. The
agency put humans on the Moon, built the International Space Station, and regularly
sends sophisticated probes billions of kilometers to Mars, Titan, and other moons and
planets with unerring precision and accuracy. All these accomplishments are achieved
under intense public scrutiny. Paradoxically, the successes of contemporary probes
such as the Mars Exploration Rover and the joint NASA–ESA Cassini–Huygens
Saturn missions have prompted some of the white suits to argue a case for a strictly
robotic paradigm which, they claim, is cheaper and holds more promise than human
exploration. In reality, nothing could be further from the truth since, although the
aforementioned missions represent a *tour de force* of exploration, robots have many
more limitations than humans. A manned lunar outpost is smart not just because the
Moon is close but because it offers a unique location from which to parameterize
human ecology on the high frontier, a goal that could not be achieved by simply

sending robots. Furthermore, the Moon is the stepping stone to an understanding of the practicalities of human survival for ever-longer periods in artificial ecosystems. Such goals can only be achieved by putting humans on the lunar surface for an extended timescale. Thanks to its cadre of superb engineers and can-do attitude NASA will take us to the Moon again. Here's how the agency will do it.

Acknowledgments

My first and greatest thanks go to my wife, Doina, for her patience and support during the writing of this book. Without Doina's exceptional grammatical talents and the endless hours she spent editing, this book could simply not have been written.

Once again, I am grateful to Clive Horwood of Praxis and the enthusiasm of the book's copy editor and typesetter, Neil Shuttlewood, is acknowledged. Likewise it has been a pleasure to work with my agent, Stephanne, and I look forward to the next book.

Along the path to writing this book, a unique group of colleagues and friends have supported my interest in manned spaceflight. My Ph.D. supervisors Professors David Grundy and Paul Enck constantly supported me in my research endeavors and have been instrumental in my being able to pursue a research career. Dr. Andrew Blaber kindly offered me a post-doctoral position at Simon Fraser University's Environmental Physiology Unit.

Finally, to those friends who read my first book and provided encouraging comments and to those friends who assured me they have the best intentions of reading my book: Julian Wigley, Tim Donovan, Gita Nand, Tania Meloni, Calvin Sandiford, Lee Williams, Tom Rodgers, Nancy Westrom, and Simba. Also, Dan Baouya—if all else fails, there is always Plan B!

Finally, thanks must go to our two cats, Mini-Mach and Jasper, and the constant welcome distraction they provided.

To my parents

About the author

Erik Seedhouse is an aerospace scientist with ambitions to become an astronaut. He experienced his first taste of micro-gravity while working as a research subject during the European Space Agency's 22nd Parabolic Flight Campaign in 1995. He gained his Ph.D. in Physiology while working at the German Space Agency's Institute for Space Medicine in Cologne between 1996 and 1998 and recently worked as an astronaut training consultant for Bigelow Aerospace in Las Vegas. He is a Fellow of the British Interplanetary Society and a member of the Aerospace Medical Association. When not writing books about space Erik flies his Cessna, races Ironman and Ultraman triathlons, climbs mountains, and spends as much time as possible in Kona and on Hapuna Beach on the Big Island of Hawaii.

Erik lives with his wife and two cats on the Niagara Escarpment in Canada.

Figures

Tables

Acronyms and abbreviations

ACES	Advanced Crew Escape Suit
ACLS	Advanced Cardiac Life Support
ACTS	Advanced Crew Transportation System
AETB-8	Alumina Enhanced Thermal Barrier-8
AIAA	American Institute of Aeronautics and Astronautics
ALARA	As low as reasonably achievable
ALHAT	Autonomous landing and hazard avoidance technology
AOD	Automatic opening device
APG	Advanced Programs Group
APMC	Agency Program Management Council
AR&D	Automated rendezvous and docking
ARC	Ames Research Center
ARPCS	Atmosphere Revitalization Pressure Control System
ARS	Acute radiation syndrome
ARS	Air Revitalization System
ASI	Artemis Society International
ASI	Augmented Spark Igniter
ASM	Aft Service Module
ATCO	Ambient temperature catalytic oxidation
ATCS	Active Thermal Control System
ATHLETE	All-Terrain-Hex-Legged Extra-Terrestrial Explorer
ATO	Abort to orbit
ATS	Aft Thrust Structure
ATSS	Advanced Transportation System studies
AUS	Advanced Upper Stage
AV	Ancillary Valve
BFO	Blood-forming organ
BLSS	Biological Life Support System

BMD	Bone mineral density
BMI	Bismaleimide
BMU	Battery Module Unit
BPC	Boost Protective Cover
BUAA	Beijing's University of Aeronautics and Astronautics
C&C	Command and control
C&N	Communications and navigation
C3I	Command, control, communication, and information
CAD	Computer-aided design
CAD	Coronary artery disease
CAIB	Columbia Accident Investigation Board
CaLV	Cargo Launch Vehicle
CAM	Computer-assisted manufacturing
CARD	Constellation Architecture Requirements Document
CAS	Chinese Academy of Sciences
CBO	Congressional Budget Office
CC	Cargo Container
CCB	Common Core Booster
CCDH	Command, control, and data handling
CDE	Carbon dioxide electrolysis
CDF	Concurrent Design Facility
CDM	Crew Descent Mission
CDMKS	Crew Descent Mission Kick Stage
CDR	Critical Design Review
CDS	Crew Descent Support
CDV	Cargo Delivery Vehicle
CE&R	Concept Exploration and Refinement (program)
CEV	Crew Exploration Vehicle
CFD	Computational fluid dynamics
CG	Center of gravity
CH_4	Methane
CHeCS	Crew Health Care System
CLL	Cargo Lunar Lander
CLV	Crew Launch Vehicle
CM	Crew Module
CMC	Center Management Council
CME	Coronal mass ejection
CMO	Crew Medical Officer
CMRS	Carbon Dioxide and Moisture Removal System
CNS	Central nervous system
CNSA	China National Space Administration
CONUS	Continental United States
COTS	Commercial Orbital Transportation System
CP	Center of pressure
CPDS	Charged Particle Directional Spectrometer

CRaTER	Cosmic Ray Telescope for the Effects of Radiation
CRC	Crew Re-Entry Capsule
CRS	Congressional Research Service
CSA	Canadian Space Agency
CSHL	Cargo Star Horizontal Lander
CSSS	Constellation Space Suit System
CTM	Crew Transfer Module
CVO	Cargo variant of Orion
CXV	Crew Transfer Vehicle
DAEZ	North Atlantic Downrange Abort Exclusion Zone
DASH	Descent Assisted Split Habitat
DAU	Data Acquisition Unit
DC-X	Delta Clipper Experimental
DCR	Design Certification Review
DCS	Decompression sickness
DDT&E	Design, development, testing, and evaluation
DIRECT	Direct Shuttle Derivative
DoD	Department of Defence
DoF	Depth of field
DOI	Descent orbit insertion
DPT	Decadel Planning Team
DRM	Design Reference Mission
DSB	Double-strand break
DS	Descent Stage
DSE-Alpha	Deep Space Exploration-Alpha
DSE	Deep-space exploration
DSM	Direct Staged Mission
DSS	Deep Space Shuttle
DSS	Deceleration Subsystem
DTA	Drop Test Article
DTE	Direct to Earth
ECLSS	Environmental Control and Life Support System
EDS	Earth Departure Stage
EELV	Evolved Expendable Launch Vehicle
EES	Emergency Egress System
EH	Escape Habitat
EIRA	ESAS Initial Reference Architecture
ELPO	Exploration Launch Projects Office
ELV	Expendable Launch Vehicle
EML1	Earth–Moon Lagrange Point 1
EMLR	Earth–Moon Lagrange rendezvous
EMS	Electronic Meeting System
EMU	Extravehicular Activity Mobility Unit
EOI	Earth orbit insertion
EOR	Earth orbit rendezvous

EOR–LOR	Earth orbit rendezvous–lunar orbit rendezvous
EPS	Electrical Power System
ERO	Earth rendezvous orbit
ESA	European Space Agency
ESAS	Exploration Systems Architecture Study
ESMD	Exploration Systems Mission Directorate
ESTEC	European Space Research and Technology Centre
ESTRACK	European Space Tracking
ET	External Tank
ETDP	Exploration Technology Development Program
ETO	Earth to orbit
EUS	Expendable Upper Stage
EVA	Extravehicular activity
FAS	Flight Analysis System
FAST	Flight application of spacecraft technologies
FBR	Fixed Base Radio
FIRST	Flight-oriented Integrated Reliability and Safety Tool
FLO	First Lunar Outpost
FOM	Figure of merit
FRR	Flight Readiness Review
FS	First Stage
FSAM	First Stage Avionics Module
FSM	Forward Service Module
FSO	Family Support Office
FSRCS	First Stage Roll Control System
FSS	Fixed Service Structure
FTI	Fusion Technology Institute
FTV	Flight Test Vehicle
GAO	Government Accountability Office
GCR	Galactic cosmic radiation
GGI	Gas generator ignition
GHe	Gaseous helium
GLOW	Gross lift-off weight
GN&C	Guidance, Navigation & Control
GOX	Gaseous oxygen
GPC	General purpose computer
GPS	Global Positioning System
GR&A	Ground rules and assumption
GRC	Glenn Research Center
GSFC	Goddard Space Flight Center
Gy	Gray
H-Suit	Hybrid Suit
HCM	Habitat Crew Module
He-3	Helium-3
HEAT	High-fidelity environment analog training

HGDS	Hazardous Gas Detection System
HHFO	Habitability and Human Factors Office
HLLV	Heavy Lift Launch Vehicle
HLM	Habitat Logistics Module
HLR	Human Lunar Return (study)
HLV	Heavy Lift Vehicle
HM	Habitation Module
HMD	Head Mounted Display
HMM	Habitat Maintenance Module
HPDE	High-density polyethylene
HPS	Human Patient Simulator
HPUC	Hydraulic Power Unit Controller
HSM	Habitat Science Module
HSSV	Helium Spin Start Valve
HSVG	Human Spaceflight Vision Group
HTPB	Hydroxyterminator polybutadiene
HUD	Heads Up Display
ICES	Integrated Cryogenic Evolved Stage
IEB	Ion Exchange Bed
ILOB	Icarus Lunar Observatory Base
IMLEO	Initial mass in low Earth orbit
INS	Inertial Navigation System
InSAR	Interferometric Synthetic Aperture Radar
IPT	Integrated Product Team
IRED	Interim Resistive Exercise Device
ISEMSI	Isolation Study for European Manned Space Infrastructure
ISP	Integrated Space Plan
I_{SP}	Specific impulse
ISRU	*In situ* resource utilization
ISS	International Space Station
ITV	Interplanetary Transfer Vehicle
IUA	Instrument Unit Avionics
IVA	Intravehicular activity
JAXA	Japan's Aerospace Exploration Agency
JCC	Jupiter Common Core
JLS	Jupiter Launch System
JPL	Jet Propulsion Laboratory
JSC	Johnson Space Center
JUS	Jupiter Upper Stage
KSC	Kennedy Space Center
L/D	Lift to drag (ratio)
L1	Lagrange Point 1
LAD	Liquid Acquisition Device
LADAR	Laser detection and ranging
LAMP	Lyman Alpha Mapping Project

LandIR	Landing and Impact Research (NASA Langley facility)
LAS	Launch Abort System
LAT	Lunar Architecture Team
LBNP	Lower-body negative pressure
LCD	Liquid crystal display
LCG	Liquid Cooling Garment
LCH_4	Liquid methane
LCROSS	Lunar Crater Observation and Sensing Satellite
LCT	Lunar Communication Terminal
LCT	Long Duration Cryogenic Tank
LEB	Lunar Exploration Base
LEM	Lunar Excursion Module
LEND	Lunar Exploration Neutron Detector
LEO	Low Earth orbit
LES	Launch Escape System
LEV	Lunar Excursion Vehicle
LExSWG	Lunar Exploration Science Working Group
LH_2	Liquid hydrogen
LIDAR	Light detection and ranging
LIDS	Low Impact Docking System
LiOH	Lithium hydroxide
LLAN	Lunar Local Area Network
LLO	Low lunar orbit
LLOX	Lunar liquid oxygen
LLPS	Lunar Lander Preparatory Study
LM	Lander Module
LM	Logistics Module
LMM	Lunar mission mode
LOC	Loss of crew
LOI	Lunar orbit insertion
LOLA	Lunar Orbiter Laser Altimeter
LOM	Loss of mission
LOR	Lunar orbit rendezvous
LOX	Liquid oxygen
LPMR	Lunar Polar Mission Rover
LPRP	Lunar Precursor Robotic Program
LRC	Langley Research Center
LRC	Lunar Resources Company
LRL	Lunar Reconnaissance Lander
LRO	Lunar Reconnaissance Orbiter
LRO	Lunar rendezvous orbit
LROC	Lunar Reconnaissance Orbiter Camera
LRS	Lunar Relay Satellite
LSAM	Lunar Surface Access Module
LSE	Lunar Surface Explorer

LSMS	Lunar Surface Mobility System
LSS	Life Support System
LTO	Lunar transfer orbit
LTV	Lunar Transfer Vehicle
LUT	Launcher Umbilical Tower
LV	Launch Vehicle
M^3	Manned Mission to the Moon
MAF	Michoud Assembly Facility
MAH	Mission Ascent Habitat
MAV	Minimum volume Ascent Vehicle
Max-ATO	Maximized abort to orbit
Max-TAL	Maximized targeted abort landing
MBR	Model-based reasoning
MCMI	Million Clinical Multiphasic Inventory
MCP	Mechanical counter-pressure
MDR	Major Design Review
MDU	Manufacturing Demonstration Unit
MECO	Main engine cut-off
MIT	Massachusetts Institute of Technology
MLAS	Max Launch Abort System
MLI	Multilayer insulation
MLP	Mobile Launcher Platform
MLUT	Minimal Launch Umbilical Tower
MM	Mission Module
MMH	Monomethyl hydrazine
MMO	Mission Management Office
MMOD	Micrometeroid/orbital debris
MMPI	Minnesota Multiphasic Personality Inventory
MPSS	Main Parachute Support System
MRR	Manufacturing Readiness Review
MSFC	Marshall Space Flight Center
MTV	Mars Transit Vehicle
NASA	National Aeronautics and Space Administration
NCRP	National Council on Radiation Protection
NEEMO	NASA Extreme Environment Mission Operations
NExT	NASA Exploration Team
NOAA	National Oceanic and Atmospheric Administration
NPR	NASA procedural requirement
NSBRI	National Space Biomedical Research Institute
NSD	NASA Standard Detonator
NTO	Nitrogen tetroxide
OBS	Operational Bioinstrumentation System
OExP	Office of Exploration
OMB	Office of Management and Budget
OML	Outer mold line

OMS	Orbital Maneuvering System
OBS	Operational Bioinstrumentation System
ORN	Osteoradionecrosis
OSC	Orbital Sciences Corporation
OSP	Orbital Space Plane
OTIS	Optimal trajectories via implicit simulation
OTV	Orbital Transfer Vehicle
P/LOC	Probability of loss of crew
P/LOM	Probability of loss of mission
PBAN	Polybutadiene acrylic acid acrylonitrile
PCA	Pneumatic control assembly
PCC	Pressurized Cargo Carrier
PCR	Pressurized Crew Rover
PCU	Power control unit
PDI	Powered descent initiation
PDR	Preliminary Design Review
PE	Polyethylene
PEG	Powered explicit guidance
PFTE	Poly-tetrafluorethylene
PICA	Phenolic-impregnated carbon ablator
PLSS	Portable Life Support System
PM	Payload Module
PMAD	Power Management and Distribution (system)
PNT	Position, navigation, and timing
POD	Point of departure
PPA	Power Pack Assembly
PPO_2	Partial pressure of oxygen
PSG	Psychological Services Group
PV	Photovoltaic
R&D	Research and development
r.m.s.	Root mean square
RATS	Research and Technology Study
RCS	Reaction Control System
RCT	Reaction Control Thruster
RDM	Robotic Descent Module
REI	Rear Entry I-Suit
REID	Risk of exposure-induced death
REM	Radiation equivalent man
RFA	Request for action
RFC	Regenerative Fuel Cell
RLV	Reusable Launch Vehicle
RM	Re-entry Module
RM	Resource Module
RMS	Remote Manipulator System
ROC	Resnick, O'Neill, Cramer

RSC	Rocket and Space Corporation
RSRB	Reusable Solid Rocket Booster
RSRM	Reusable Solid Rocket Motor
RSS	Rotating Service Structure
S&MA	Safety Mission Assurance Office
SAEH	Support Ascent Escape Habitat
SAGES	Shuttle and Apollo Generation Expert Services
SAR	Synthetic Aperture Radar
SARSAT	Search and rescue satellite-aided tracking
SBIR	Small Business Innovative Research
SCA	Spacecraft Adapter
SCR	Solar cosmic ray
SEI	Space Exploration Initiative
SEM	Space Exploration Module
SF	Factor of safety
SFINCSS	Simulation of Flight of International Crew on Space Station
SH	Surface Habitat
SLS	Saturn Launch System
SM	Service Module
SOHO	Solar and Heliospheric Observatory
SORT	Simulation and optimization of rocket trajectories
SPACE	Screening Program for Architecture Capability Evaluation
SPE	Solar particle event
SPM	Surface Power Module
SPWE	Solid Polymer Water Electrolysis
SQM	Strange quark matter
SRB	Solid Rocket Booster
SRM	Solid Rocket Motor
SRR	System Requirements Review
SS	Satellite and storage
SSB	Single-strand break
SSC	Stennis Space Center
SSC	Systems and Software Consortium
SSME	Space Shuttle Main Engine
SSRB	Space Shuttle Solid Rocket Booster
SS	Satellite and storage
SSTO	Single stage to orbit
Sv	Sievert
SYZ	Soyuz
TAL	Targeted abort landing
TEI	Trans-Earth injection
TEPC	Tissue Equivalent Proportional Counter
TIM	Technical Interface Meeting
TLI	Translunar injection
TO	Thrust oscillation

TPI	Terminal phase initiation
TPS	Thermal Protection System
TRL	Technology readiness level
TT&C	Telemetry, tracking, and control
TUR	Trencher Utility Rover
UAS	Untargeted abort splashdown
UCR	Unpressurized Crewed Rover
UDMH	Unsymmetrical dimethylhydrazine
UHF	Ultrahigh frequency
US	Upper Stage
USAF	United States Air Force
USEE	Upper Stage Engine Element
USEO	Upper Stage Element Office
UV	Ultraviolet
VAB	Vehicle Assembly Building
VCC	Voice call continuity
VDC	Volts Direct Current
VEG	Virtual Environment Generator
VPCAR	Vapor Phase Catalytic Ammonia Removal (system)
VR	Virtual reality
VSE	Vision for Space Exploration
VT	Ventricular tachycardia
WFRD	Wiped Film Rotating Disk
YPG	Yuma Proving Grounds

1

Vision for Space Exploration

As for the future, your task is not to see it, but to enable it.
Antoine de-Saint Exupéry (1900–1944)
The Wisdom of the Sands

America has not developed a new vehicle to advance human exploration in space in nearly a quarter of a century. It is time for America to take the next steps. Today I announce a new plan to explore space and extend a human presence across our solar system. We will begin the effort quickly, using existing programs and personnel. We'll make steady progress, one mission, one voyage, one landing at a time.
President George W. Bush
speaking at NASA HQ, January 14th, 2004

The above excerpt from President Bush's speech launched a bold and forward-thinking space exploration policy now known as the Vision for Space Exploration (VSE) that set a direction for NASA's human spaceflight program for decades to come. To accomplish the VSE, President Bush assigned NASA to initiate the Constellation Program, tasked with developing a new Crew Exploration Vehicle (CEV) and a new heavy-lift vehicle. For space enthusiasts and those working in the space industry, the VSE announcement was a statement long overdue.

PRE-VSE SPACE EXPLORATION INITIATIVES

President George W. Bush's January 2004 speech echoed some of the objectives established on July 20, 1989, when his father, President H.W. Bush, announced a new Space Exploration Initiative (SEI), which directed NASA to conduct a 90-day study to define mission architectures intended to establish a lunar outpost as a test-bed for expeditions to Mars. Richard Darman, then Director of the Office of Management and Budget (OMB), estimated fulfilling the SEI goal of landing on the Moon and Mars would cost $400 billion. Predictably, the project sank like the proverbial stone, leaving space enthusiasts less than impressed. Fifteen years later, no mention was made of the anticipated cost of returning humans to the Moon, although a Congressional Research Service (CRS) report estimated NASA will need $104 billion. Despite the projected cost, surveys have shown a majority of the public support the VSE, as evidenced by an Associated Press–Ipsos poll of the public's initial reaction to the President's speech, which found 48% of respondents in favor of a return to the Moon [1]. A later poll conducted by Gallup between June 22 and July 7, 2004, found that, providing NASA's budget did not exceed 1% of the federal budget, 26% strongly supported the program, 42% supported, 15% opposed, and 9% strongly opposed the plan.

THE VISION FOR SPACE EXPLORATION PLAN

As summarised in Table 1.1, NASA's VSE plan for exploring space and sending astronauts to the Moon and beyond is bold and far-reaching. In fact, the stated goal of ultimately conducting manned missions to Mars and beyond has caused some observers to suggest such a plan may be over-ambitious. However, the means to achieve the objectives and bringing the vision to reality are carefully articulated in documents describing the exploration building blocks that comprise the Constellation Program, an overview of which is described in this chapter.

The first goal: exploration activities in low Earth orbit

"To meet this goal, we will return the space shuttle to flight as soon as possible, consistent with safety concerns and the recommendations of the Columbia Accident Investigation Board. The shuttle's chief purpose over the next several years will be to help finish assembly of the International Space Station. In 2010, the space shuttle, after nearly 30 years of duty, will be retired from service."

As this book is being written, NASA is pressing ahead with the design and development of a new spaceship and family of launch vehicles (LVs) that will replace the Space Shuttle, which is due to retire on September 30th, 2010. Replacing the Shuttle will be no easy task. Programs such as the X-33 and X-38 were designed to replace the Shuttle with an elegant single-stage-to-orbit (SSTO) vehicle but cost overruns and insurmountable technological challenges resulted in both programs being canceled. By using mostly Shuttle-derived technology to build its new family of LVs, NASA hopes to avoid the expensive mistakes of the X-33 and X-38. Nevertheless, the design of two new LVs, complete with subsystems and launch infrastructure, will be challenging.

Table 1.1. Synopsis of NASA's plan for space exploration.

Complete the International Space Station
Safely fly the Space Shuttle until 2010
Develop and fly the Orion Crew Exploration Vehicle no later than 2015
Return to the Moon no later than 2020
Extend human presence across the solar system and beyond
Implement a sustained and affordable human and robotic program
Develop supporting innovative technologies, knowledge, and infrastructures
Promote international and commercial participation in exploration
Use the Moon to prepare for future human missions to Mars and other destinations
Pursue scientific activities to address fundamental questions about the solar system, the universe, and our place in them
Extend sustained human presence to the Moon, to enable eventual settlement
Expand Earth's economic sphere to encompass the Moon and pursue lunar activities with direct benefits to life on Earth
Strengthen existing and create new global partnerships
Engage, inspire, and educate the public

The first of the new LVs is called Ares I (Figure 1.1) and is the designated Crew Launch Vehicle (CLV). The two-stage Ares I, which will feature a launch escape system (LES), a five-segmented booster, and powerful J-2X engine, will carry the CEV into low Earth orbit (LEO) and is scheduled for its first unmanned test launch in April 2009. Once the proving flights of Ares I are over, NASA will begin to prepare for returning astronauts to the Moon. For this they will need the services of a heavy-lift LV, which will carry into orbit hardware such as the Earth Departure Stage (EDS) which will provide the burn sending the CEV beyond LEO. The huge rocket, named Ares V (Figure 1.2), will look like a hybrid of the Space Shuttle and the Saturn V. Powered by five mighty J-2X engines attached to the base of the first-stage (FS) external tank (ET), Ares V's journey to LEO will be assisted by a pair of five-segmented SRBs bolted onto the ET and an upper stage (US) powered by J-2S engines originally used during the Apollo Program in the 1960s.

The second goal: space exploration beyond low Earth orbit

"Our second goal is to develop and test a new spacecraft, the crew exploration vehicle, by 2008 and to conduct the first manned mission no later than 2014. The

Figure 1.1. NASA's Crew Launch Vehicle, Ares I, will carry the Orion Crew Exploration Vehicle. Image courtesy: NASA.

crew exploration vehicle will be capable of ferrying astronauts and scientists to the space station after the shuttle is retired. But the main purpose of this spacecraft will be to carry astronauts beyond our orbit to other worlds. This will be the first spacecraft of its kind since the Apollo command module.''

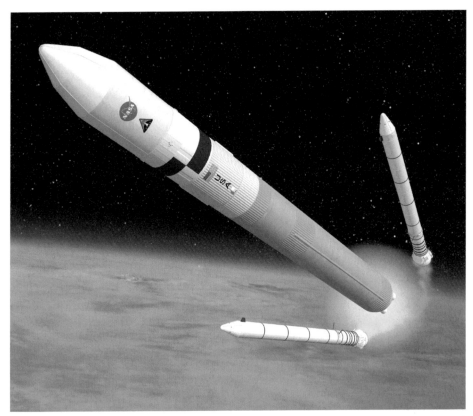

Figure 1.2. NASA's Cargo Launch Vehicle, Ares V, will carry the Earth Departure Stage and Lunar Surface Access Module into low Earth orbit. Image courtesy: NASA.

The CEV, in which astronauts will fly to and from LEO and to and from the Moon, is called Orion (Figure 1.3, see color section). It will be the same shape as the Apollo command module but will provide astronauts with three times as much habitable volume and, unlike its illustrious predecessor, will be designed to be re-used for up to ten times. In space, Orion will have a liquid-oxygen-and-liquid-methane-powered Service Module (SM) attached that will soak up energy from its pair of electricity-generating solar arrays.

The third goal: space transportation capabilities supporting exploration

"Our third goal is to return to the Moon by 2020, as the launching point for missions beyond. Beginning no later than 2008, we will send a series of robotic missions to the lunar surface to research and prepare for future exploration. Using the crew exploration vehicle, we will undertake extended human missions to the Moon as early as 2015, with the goal of living and working there for increasingly extended periods of time. Eugene Cernan, who is with us today, is the last man to set foot on the lunar surface. He said this as he left: 'We leave as we came and, God willing, as we shall return, with peace, and hope for all mankind.' America will make those words come true.''

The month following President Bush's address, NASA Administrator, Sean O'Keefe, released the VSE [4], NASA's roadmap of the agency's many programs that support the vision laid out in the speech. To develop the many technologies and systems required to realize the VSE, the Exploration Systems Mission Directorate (ESMD) was created. The following year, in April 2005, NASA's new Administrator, Dr. Michael Griffin, commissioned the Exploration Systems Architecture Study (ESAS), which evaluated the ways NASA could realize the VSE. The ESAS represents one of the primary building blocks of the Constellation Program as it documents how the VSE will be achieved.

EXPLORATION SYSTEMS ARCHITECTURE STUDY CHARTER

The ESAS began on May 2nd, 2005 and was assigned with performing four tasks by July 29th, 2005. It was comprised of 20 core team members, collocated at NASA HQ and hundreds of NASA employees from various NASA centers. To direct the ESAS team a number of ground rules and assumptions (GR&As) were established. The GR&As included guidelines and constraints that were directed at factors such as safety, operations, technical, cost, schedule, testing, and foreign assets, an overview of which is outlined here.

Exploration Systems Architecture Study ground rules and assumptions

The Safety and Mission Assurance GR&As that the ESAS team was guided by was a NASA Procedural Requirement (NPR) document known as the Human-Rating Requirements for Space Systems document (see Glossary). This set of requirements was used as a reference when the ESAS team defined mission aspects such as abort opportunities and orbital operations.

Notable Operations GR&As included requirements to deliver crew and cargo to and from the ISS until 2016, to devise an architecture that separated crew and large cargo to the maximum extent possible and for all on-orbit flight operations to be conducted at the Kennedy Space Center (KSC).

Examples of some of the Technical GR&As included restrictions requiring the CEV to be designed for a crew of four for lunar missions and a crew of six for ISS missions. The ESAS team were also required to follow certain factors of safety for elements such as the crew cabin, new and redesigned structures, and margins for rendezvous and docking operations.

One of the Schedule GR&As was a requirement for the first CEV flight to the ISS by 2011 and to perform the first lunar landing by 2020. Since the publication of the

ESAS study the first of these requirements has slipped and it is uncertain how much this slip will impact the lunar landing date.

Flight-testing of the new LVs will be constrained by the Testing GR&As which require that qualification of the CEV will demand that at least one fully functional flight be completed prior to the CEV being human-rated. Similarly, before the new Ares I LV can be human-rated, at least three flight tests must be conducted.

Exploration Systems Architecture Study tasks

Some of the first tasks of the ESAS team included an assessment of the CEV requirements, defining how the CEV could provide crew transport to the ISS and devising a plan to reduce the time between the Shuttle's retirement and the first manned flights of the CEV. Next, the ESAS defined the requirements and configurations for the LV systems to support manned missions to the Moon and Mars. The ESAS then developed reference lunar architectures for manned missions to the Moon and identified the technologies required to enable the architectures. These four tasks were transposed into the four major points of the ESAS Charter, which were definition of the CEV, definition of the LVs, definition of the lunar architecture, and definition of the technology required.

To fulfil their tasks the ESAS effort performed hundreds of trade studies that examined options ranging from an assessment of different CEV shapes to determining optimum EVA requirements. These trade studies also quantitatively and qualitatively assessed costs, schedule, reliability, safety, and risk associated with all factors associated with the elements of the architecture required to return humans to the surface of the Moon. To more accurately define the elements required for the architecture the ESAS team also utilized Figures of Merit (FOM), a disciplined, well-proven method of characterizing the performance of a component relative to other components of the same type (Table 1.2).

Mission architecture overview

Once the ESAS team had evaluated the FOMs they proceeded with devising a series of Design Reference Missions (DRMs) that defined how crews would be transported to and from the ISS and how pressurized and unpressurized cargo would be transported to and from the ISS. The evolution of the DRMs is described in Chapter 7. The DRMs, an overview of which is described here, also defined how crew and cargo would be transported to and from the lunar surface and how crew and cargo would be transported to and from an outpost at the lunar South Pole.

Design Reference Missions overview

DRM: Transportation of astronauts to and from the ISS

The objective of this DRM is to ferry crews of three astronauts to the ISS where theywill complete six-month mission increments. During the mission (Figure 1.4)

Table 1.2. Exploration Systems Architecture Study figures of merit.

Safety and mission success	Effectiveness performance	Extensibility flexibility	Programmatic risk	Affordability
Probability of loss of crew (P/LOC)	Cargo delivered to lunar surface	Lunar mission flexibility	Technology development risk	Technology development cost
Probability of loss of mission (P/LOM)	Cargo returned from lunar surface	Extensibility to other exploration destinations	Cost risk	Design, development, test, and evaluation cost
	Surface accessibility	Commercial extensibility	Schedule risk	Facilities cost
	Usable surface crew-hours	National security extensibility	Political risk	Operations cost
	System availability			Cost of failure
	System operability			

crewmembers will be provided with an anytime return capability made available by an Orion berthed at the ISS.

Orion will be launched by an Ares I into LEO at an inclination of 51.60, after which a series of burns will be performed designed to raise the orbit of Orion so it eventually catches up with the ISS. Once Orion is close to the ISS, a standard rendezvous and docking maneuver will be performed, the crew will ingress, and

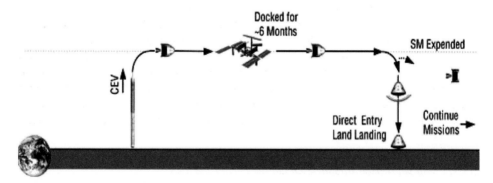

Figure 1.4. Design Reference Mission showing how astronauts will be transported to and from the International Space Station. Image courtesy: NASA.

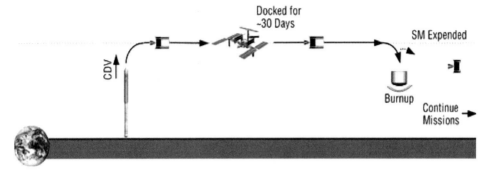

Figure 1.5. Design Reference Mission showing how unpressurized cargo will be transported to and from the International Space Station. Image courtesy: NASA.

Orion will assume the role of "rescue vehicle" for the duration of the mission increment. Upon completion of the mission, the SM will be discarded, Orion will execute a de-orbit burn and then perform a nominal landing either on land or water.

DRM: Transportation of unpressurized cargo to the ISS

The objective of this mission will be simply to haul unpressurized cargo to the ISS, using a Cargo Delivery Vehicle (CDV) and an Ares I (Figure 1.5). Following launch into LEO at an inclination of 51.6°, the CDV will perform a series of orbit-raising burns to close with the ISS before conducting a standard onboard-guided approach designed to place it within reach of the ISS's Remote Manipulator System (RMS). The ISS crew will then use the RMS to grapple the CDV and berth it to the ISS, where it will remain for 30 days before it will be unberthed, and conduct a ground-validated de-orbit burn for a destructive re-entry.

DRM: Transportation of pressurized cargo to and from the ISS

The purpose of this mission will be to ship pressurized cargo to and from the ISS and return to Earth following a 90-day berthing period at the ISS (Figure 1.6). The mission profile will utilize a cargo variant of Orion (CVO), loaded with up to 3,500 kg of logistics, and an Ares I. Following insertion into LEO the CVO will perform orbit-raising burns to chase the ISS. Once at a safe station-keeping position relative to the ISS, Mission Control will command the CVO to perform an onboard-guided approach, the CVO will dock, pressurization checks will be conducted, and ingress of cargo will be performed by the ISS crew. Following a 90-day docking phase the CVO will perform a de-orbit burn and perform a nominal landing either on land or water.

DRM: Manned lunar mission with cargo

The lunar sortie mission (Figure 1.7) is designed to transport four astronauts to any site on the Moon, a capability referred to as *global access*. Once there, the astronauts

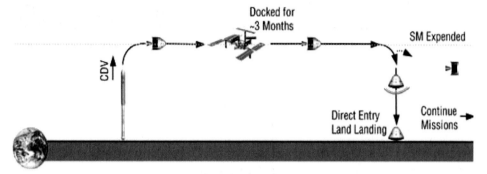

Figure 1.6. Design Reference Mission showing how pressurized cargo will be transported to and from the International Space Station. Image courtesy: NASA.

will spend up to seven days performing scientific and exploration tasks in pairs. A lunar sortie mission will require an Ares I, an Ares V, an Orion, a Lunar Surface Access Module (LSAM), and an EDS.

The mission will commence with the launch of the Ares V, which will deliver the LSAM and EDS to LEO. The Ares I will then deliver Orion to LEO, where Orion will rendezvous and dock with the LSAM–EDS configuration. The EDS will perform a

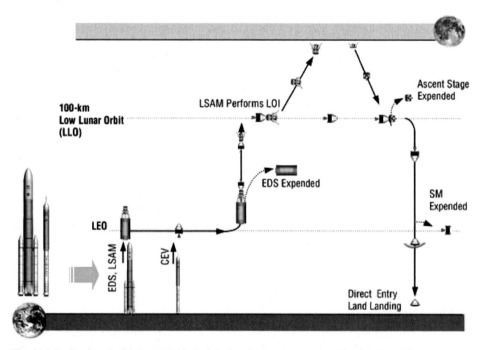

Figure 1.7. Design Reference Mission showing how astronauts will travel to the Moon with cargo. Image courtesy: NASA.

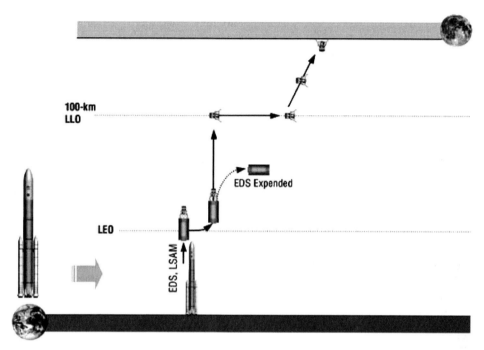

Figure 1.8. Design Reference Mission showing how cargo will be transported to the lunar surface. Image courtesy: NASA.

trans-lunar insertion (TLI) burn and then be discarded. The LSAM will then perform a burn that will insert the LSAM and Orion into lunar orbit. Once in lunar orbit, all four astronauts will transfer to the LSAM, undock from Orion, and descend to the lunar surface. After spending seven days exploring the lunar surface the LSAM will perform an ascent to low lunar orbit (LLO) and rendezvous and dock with Orion, which will return to Earth for a land or water landing.

DRM: Cargo transportation to lunar surface

This mission will be utilized to help establish a permanent human presence on the Moon by ferrying up to 20 tonnes of cargo to the lunar surface (Figure 1.8). To achieve this mission Ares V will launch an EDS and cargo variant of the LSAM into LEO, the EDS will perform the TLI burn, and the LSAM will perform the burn to insert itself into lunar orbit.

DRM: Outpost mission with crew and cargo

To support a six-month surface increment on the lunar surface, this DRM will transfer a crew of four astronauts and supplies in a single mission using the suite of vehicles shown in Figure 1.9. First, the LSAM and EDS will be deployed following a single Ares V launch into LEO. Then, an Ares I will launch the Orion into LEO and

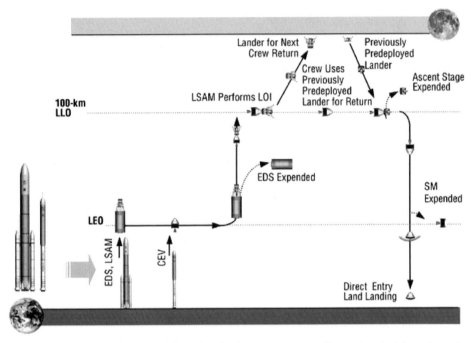

Figure 1.9. Design Reference Mission showing how astronauts will travel to the Moon for a six-month outpost mission. Image courtesy: NASA.

will rendezvous and dock with the LSAM–EDS configuration. After performing the TLI burn, the EDS will be discarded and the LSAM will perform the lunar orbit insertion (LOI) burn for Orion and the LSAM. Once in lunar orbit, the crew will transfer to the LSAM, descend to the lunar surface, and commence a six-month increment on the surface of the Moon, at the end of which the LSAM will return the crew to lunar orbit where the LSAM will dock with Orion and the crew will transfer to the Orion for the journey back to Earth, leaving the LSAM to impact the lunar surface.

The DRMs devised by the ESAS represented a time-phased evolutionary architecture designed to return humans to the surface of the Moon. The details of the DRMs and other elements of the architecture such as the design of Orion and Ares I was published in the ESAS report. Shortly after the report's publication, on December 30th, 2005, NASA's plans for realizing the vision became law with the passing of the NASA Authorization Act of 2005.

> "The Administrator shall establish a program to develop a sustained human presence on the moon, including a robust precursor program to promote exploration, science, commerce and U.S. pre-eminence in space, and as a stepping stone to future exploration of Mars and other destinations."
>
> NASA Authorization Act of 2005

THE CONSTELLATION PROGRAM

To achieve the goals set by the VSE, the Constellation Program (Figure 1.10) is leveraging expertise across NASA (Table 1.3), a strategy that is gradually transforming the centers into a unified agency-wide team dedicated to returning humans to the Moon.

Table 1.3 provides some sense of the daunting scale of the Constellation Program. Although the continued operations of the Space Shuttle constrain the budget and workforce available to develop Constellation through 2010, NASA is making the best use of its experienced employees through the phased retirement of the Shuttle. Already, targeted activities in lunar capability development have begun, initial concept development of the lunar lander and Ares V are underway, and a schedule of the major vehicle engine tests and missions has been published (Table 1.4).

The Space Shuttle and ISS programs were developed in the context of their times, facing significant unknown technical and operational challenges. Similarly, NASA leadership has given the Constellation Program license to explore new methods and approaches to develop and procure the launch vehicles and spacecraft necessary to return humans to the Moon. Consequently, the elements of the Constellation Program described in Tables 1.3 and 1.4 comprise the baseline of a robust system, based on the very best of NASA's heritage.

Figure 1.10. NASA's Constellation logo. Image courtesy: NASA.

Table 1.3. NASA centers supporting the Constellation Program.

Facility name	Use of facility
Ames Research Center	
Transonic Wind Tunnel	Ares scale model testing
ARC Jet Lab	Orion components and Thermal Protection System testing
Unitary Plan Wind Tunnel	Orion components and Thermal Protection System testing
Glenn Research Center (Lewis Field)	
Instrument Research Lab	Miniature sensor and associated validation software development for leak detection
Supersonic Wind Tunnel Office and Control Building	Integrated design analysis and independent verification and validation in support of Orion vehicle design
Glenn Research Center (Plum Brook Station)	
Spacecraft Propulsion Research Facility	Earth Departure Stage testing
Space Power Facility	Orion acoustic/random vibration, thermal vacuum testing
Johnson Space Center	
Crew Systems Lab	Component and small-unit bench-top testing
Crew Systems Lab (8 ft chamber)	Uncrewed integrated EVA life support system operational vacuum-testing
Crew Systems Lab (11 ft chamber)	Crewed EVA system vacuum-testing
Crew Systems Lab (Thermal Vacuum Glovebox)	Thermal vacuum-testing of gloves and small tools
Communications & Tracking Development Lab Building	Orion test and verification
Mission Control Center	Mission control activities
Jake Garn Simulator & Training Facility	Astronaut training
Systems Integration Facility	Astronaut training
Sonny Carter Training Facility	Astronaut training
Space Environment Simulation Lab	Crewed thermal vacuum-testing

Facility name	Use of facility
Kennedy Space Center	
Launch Complex-39, Pads A and B	Ares launch facilities
SRB assembly and refurbishment facilities	Recovery and refurbishment of Ares I and Ares V vehicle elements
Missile Crawler Transporter Facilities	Crawlers used to transport Ares I and Ares V from VAB to launch pad
Crawlerway	Roadbed used by crawlers to transport Ares I and Ares V between VAB and launch pads
Mobile Launcher Platform	Transport Ares V launch vehicle from VAB to launch pad
Mobile Launcher	Platform used to transport Ares I launch vehicles from VAB to launch pad
Lightning Protection System	Launch vehicle lightning strike protection
Launch Control Center	Launch control
Vehicle Assembly Building	Vehicle assembly and integration
Operations and Checkout Building	Orion assembly and integration
Space Station Processing Facility	Candidate facility for processing of Lunar Lander
Hazardous Processing Facility	Process hazardous materials for Crew Module and Service Module prior to integration with Launch Vehicle
Orbiter Processing Facilities	Ares V Core Stage assembly
VAB Turning Basin Docking Facility	Perform maintenance activities to ensure structural and operational integrity
Parachute Refurbishment Facility	Process and refurbish parachutes for SRB and Orion operations

(continued)

Table 1.3 (*cont.*)

Facility name	Use of facility
Langley Research Center	
Materials Research Lab	Testing of materials and test components for Ares and Orion
Structures & Materials Lab	Testing of materials and test components for Orion and Ares
Thermal Lab	Stress-testing for Orion
Fabrication & Metals Technology Development Lab	Fabrication of models and test items for Orion and Ares
Wind Tunnel	Scale model testing for Orion
Mach-10 Tunnel	Scale model testing for Orion
Vertical Spin Tunnel	Scale model testing for Orion, including Launch Abort System
Transonic Dynamics Tunnel	Scale model wind tunnel testing for Orion and Ares
Gas Dynamics Complex (Mach-6 Tunnel)	Scale model wind tunnel testing for Orion and Ares
Impact Dynamics Facility	Orion drop tests
Michoud Assembly Facility	
Manufacturing Building	Ares I Upper Stage structural welding, avionics, and common bulkhead assembly
Vertical Assembly Facility	Ares I Upper Stage and Orion Crew Module, Service Module, back shell and heat shield fabrication
Acceptance & Preparation Building	Ares I Upper Stage
Pneumatic Test Facility & Control Building	Pressure and dynamic testing
High Bay Addition	Ares I Upper Stage and Ares V Core Stage assembly and foam application

Facility name	Use of facility
Marshall Space Flight Center	
Hardware Simulation Lab	Ares Upper Stage engine control system and software testing and avionics and systems integration
Avionics Systems Testbed	Ares Upper Stage avionics integration
Test Facility	Ares Upper Stage component testing
Structural Dynamic Test Facility	Ares I and Ares V ground vibration testing
Hot Gas Test Facility	Ares I First Stage design configuration certification and Upper Stage hot gas testing
Propulsion & Structural Test Facility	Testing Ares I First Stage and Ares Upper Stage pressure vessel components
Materials & Processes Lab	Materials testing
Test & Data Recording Facility	Ares Upper Stage spark igniter testing
Structures & Mechanics Lab	Ares Upper Stage engine vibration testing
Huntsville Operations Support Center	Engineering support for Ares Upper Stage development operations
Advanced Engines Test Facility	Ares Upper Stage engine testing
Multi-purpose High Bay & Neutral Buoyancy Simulator Complex	Ares Upper Stage fabrication
National Center for Advanced Manufacturing	Ares Upper Stage support actions and evaluations
Engineering & Development Lab	Final assembly and preparation for Ares Upper Stage testing
Wind Tunnel Facility	Ares wind tunnel testing
Cryogenic Structural Test Facility	Ares Upper Stage structural load tests including cryogenic testing of common bulkhead
Stennis Space Center	
Rocket Propulsion Test Stand	Ares I J-2X power pack and J-2X Upper Stage engine testing
B-1 Test Stand	Ares V RS-68B engine testing
A-3 Test Stand Vacuum Facility	Ares Upper Stage testing
B-2 Test Stand	Ares V RS-68B Core Stage engine testing

Table 1.4. Major engine tests, flight tests, and initial Constellation Program missions.

Test flight[a]	Location	Year	Estimated # of tests/flights
First Stage ground tests			
Development Motor 1 hot-fire test		2008	1
Development Motor 1 hot-fire test	ATK Promontory, Utah	2009	1
Qualification Motor hot-fire test		2011	2
Qualification Motor hot-fire test		2012	1
Launch Abort System tests			
Pad abort flight test		2008	1
Launch abort flight test		2009	1
Pad abort test	White Sands Missile Range	2010	1
Launch abort flight test		2010	1
Launch abort flight test		2011	2
Upper Stage engine (J-2X) ground tests			
Upper Stage engine hot-fire test	Stennis Space Center	2010–2014	175
Upper Stage engine hot-fire test (simulated altitude)		2010–2014	100
Upper Stage Engine hot-fire test	Glenn Research Center	2011	2
Main Propulsion Test Article hot-fire test	Marshall Space Flight Center	2010–2013	24
Ares I flights			
Ares I ascent development flight test		2009	2
Ares I ascent development flight test		2012	1
Orbital flight test	Kennedy Space Center	2013	2
Orbital flight test[b]		2014	2
Mission flight[c]		2015–2020	Up to 30

Test flight[a]	Location	Year	Estimated # of tests/flights
Ares V Core Stage engine ground tests			
RS-68B engine hot-fire test	Stennis Space Center	2012–2018	160
Main Propulsion Test Article Cluster hot-fire test		2012–2018	20
Earth Departure Stage engine ground test			
Upper Stage engine hot-fire test (simulated altitude)	Glenn Research Center	2012–2014	20
Main Propulsion Test Article hot-fire test	Marshall Space Flight Center	2015-2018	20
Ares V flights			
Flight test	Kennedy Space Center	2018	2
Mission flight[d]		2019	2
Mission flight		2020	1

[a] The number, location, and types of tests are subject to change as Constellation test programs evolve.
[b] Third orbital flight test will be the first crewed launch of Orion/Ares I.
[c] Up to five Ares I flights per year will occur.
[d] Second flight in 2019 is the first planned to include landing a crew on the Moon.

Within the structure of NASA the Constellation Program [3] is designed to evolve as near-term technical and programmatic objectives are realized, all the while remaining attentive to the lessons learned from current and prior programs. An example of the Constellation Program learning from previous programs is the creation of the Shuttle and Apollo Generation Expert Services (SAGES), a pathway enabling the enlistment of retired experts from NASA's past such as legendary Flight Directors Chris Kraft and Glynn Lunney.

The Constellation Program was staffed with recognized leadership within NASA capable of aligning the integrated set of requirements that will unite the agency's centers into a strong program able to integrate and execute the many tasks required to fulfill the objective of returning to the Moon.

RATIONALE FOR RETURNING TO THE MOON

"A large portion of the scientific community in the U.S, also prefers Mars over the Moon. But interest in the Moon is driven by goals in addition to and beyond the requirements of the science community. It is driven by the imperatives that ensue from a commitment to become a space-faring society, not primarily by scientific objectives, though such objectives do indeed constitute a part of the overall rationale."

NASA Administrator, Dr. Michael Griffin

Although the power of the VSE is politically driven, the reasons for returning to the Moon, often referred to by mission planners as *drivers*, include science, technology, exploration, and exploitation.

Science

The potential of the Moon as a scientific outpost can be divided into the categories of science *of* the Moon, science *from* the Moon, and science *on* the Moon. Science *of* the Moon includes the potential for conducting geophysical, geochemical, and geological research, which will enable a better understanding of the origin and evolution of the Moon. Science *from* the Moon includes research that can be performed as a result of the unique properties of the lunar surface, such as astronomical observations on the far side of the Moon. Science *on* the Moon includes the disciplines of biology and exobiology that investigate the stability of biological and organic systems in hostile environments, and the regulation of autonomous ecosystems. Each of these scientific disciplines can benefit from the presence of humans on the Moon.

Technology

The hostile environment of the Moon, such as the radiation field, reduced gravity, and the ever-prevalent dust, are very similar to the conditions on Mars and therefore offer a suitable test-bed to apply and evaluate technologies designed to deal with such an environment. Also of particular importance for future Mars missions is the testing of *in situ* resource utilization (ISRU) on the lunar surface, since this process will be used extensively by astronauts during Martian missions.

Other scientific objectives that must be met before undertaking manned missions to Mars include the development of autonomous tools and integrated advanced sensing systems such as bio-diagnostics, telemedicine, and environmental monitoring and control. Once again, the lunar surface offers a demanding testing ground for the rigorous evaluation and development of such systems.

Exploration and exploitation

More than two thousand people have set foot on Mount Everest, hundreds of adventurers have reached both North and South Poles and almost five hundred astronauts, cosmonauts, taikonauts, and spaceflight participants have experienced microgravity. New frontiers are opened up for the purpose of understanding the unexplored, and the next logical step in exploration is to venture beyond LEO and see the sights on our nearest neighbors.

Once a permanent base has been established on the Moon and relatively inexpensive access is available, space tourism will become a major lunar business. Given the resources of the Moon it will not be long before other industries develop infrastructures to extract life support consumables, propellant, and helium-3.

"We know from earliest recorded history, some 5,000 years ago, that human beings have always sought to learn more about their world. The century just ended has witnessed stunning advancements in science and medicine, including the launching of space exploration. There is a danger, however, that the new century may usher in an age of timidity, in which fear of risks and the obsession with cost-benefit analysis will dull the spirit of creativity and the sense of adventure from which new knowledge springs."

M.E. DeBakey (2000)

Although each of the drivers mentioned represents a strong case for returning to the Moon, an equally strong justification can be made on the grounds that space is part of our culture. Although many in the space industry complain that nobody knows the names of astronauts, that nobody cares what happens on the ISS, and that nobody gets excited by the launch of the Space Shuttle, the fact is that when the Shuttle *Columbia* broke up on re-entry on that fateful day in February 2003, the United States and much of the world came to a standstill. The national mourning that followed *Columbia* was indicative of the behavior of a nation that is not only still deeply interested in the challenge of manned spaceflight but also committed to the vision of space exploration.

In its August 2003 report, the Columbia Accident Investigation Board (CAIB) remarked that following the landing on the Moon by Apollo 11, "President Nixon rejected NASA's sweeping vision for a post-Apollo effort that involved full development of low-Earth orbit, permanent outposts on the Moon, and initial journeys to Mars" [2]. As a result of Nixon's decision, there was a "lack, over the past three decades, of any national mandate providing NASA with a compelling mission requiring human presence in space" [2]. Without such a mandate, NASA became "an organization straining to do too much with too little" [2].

The reaction to what the CAIB referred to as a "failure of national leadership" was President Bush's VSE, an ambitious and vigorous program intended to reverse a period of drought in which dreams had been limited and little investment had been made in building public or political support for space initiatives. As a result of the VSE, NASA and the United States now have an opportunity to ensure space leadership for decades to come, and once again dare to do the hard things that other nations do not.

"The human imperative to explore and settle new lands will be satisfied, by others if not by us. Humans will explore the Moon, Mars, and beyond. It's simply a matter of which humans, when, what values they will hold, and what languages they will speak, what cultures they will spread. What the United States gains from a robust program of human space exploration is the opportunity to carry the principles and values of western philosophy and culture along on the absolutely inevitable outward migration of humanity into the solar system and, eventually, beyond. These benefits are tangible and consequential. It matters what the United States chooses to do, or not to do, in space."

NASA Administrator, Michael Griffin, AIAA speech

VSE is both a rocket project and a plan for establishing the first outpost of human-kind off the Earth. Not only does the plan have the potential to make future space operations cheaper and safer, it also offers the possibility of bringing spaceflight closer to becoming routine and provides an opportunity to bring about sustainable exploration and development. VSE has the potential to create a positive future for civilization and in doing so remind humankind that the future is not limited to this lonely rock in space that many people fear is the only future we have.

"Mankind is drawn to the heavens for the same reasons we were once drawn into unknown lands and across the open sea. We choose to explore space because doing so improves our lives and lifts our national spirit. So let us continue the journey. May God bless."

President George W. Bush

REFERENCES

[1] "AP Poll: U.S. Tepid on Bush's Space Plans," *Associated Press* (January 12, 2004), 14:50.
[2] *Columbia Accident Investigation Board: NASA*. Government Printing Office, Washington, D.C. (August 2003).
[3] NASA. *Constellation Architecture Requirements Document*, Document CxP 70000, p. 20. NASA, Washington, D.C. (December 21, 2006).
[4] Smith, M.S. *Space Exploration: Issues: Concerning the "Vision for Space Exploration"*. Congressional Research Service, Washington, D.C. (January 4, 2006).

2

Racing to the Moon

Don't tell me that man doesn't belong out there. Man belongs wherever he wants to go, and he'll do plenty well when he gets there.

Wernher von Braun, *Time* magazine, 1958

The strategic importance of a manned presence in space cannot be overlooked, so it is not surprising other countries are planning to join the United States in embarking upon manned missions to the Moon.

Recognizing that the defence doctrines of the United States and Russia have always been space-oriented, China appears to be adopting a similar space strategy. While China's space endeavor started much later than the American or Russian programs, their recent forays into low Earth orbit (LEO) have demonstrated they are making rapid progress in the direction of more ambitious missions that may include establishing their own space station and landing on the Moon.

The Russians, long-time veterans of sending humans into space, have made no secret of their lunar ambitions and are planning their own program designed to land their cosmonauts on the surface of the Moon.

Competing with the Americans, Russians, and the Chinese, is the European Space Agency (ESA), whose space program is rooted more in symbolism than military might. European nations have struggled for many years to style a coherent political union that wields influence on the world stage on a par with the United States. Although the European Union's disparate nations have succeeded in competing economically they have fallen short in finding a common ground politically. To think of themselves as a single people may require a shared and inspirational goal such as putting a European on the surface of the Moon or Mars, using a spacecraft bearing the logo of the European Union.

Painting China, Russia, and the ESA as the United States' opponents in a new space race may be interpreted by some as a NASA attempt to secure more funding from Congress. While this may be true, there is no denying national rivalries exist and space is the new arena of competition, since not only does the high frontier allow nations to gain international kudos, space is also becoming increasingly vital to national security.

CHINA'S LUNAR AMBITIONS

In October 2000, China National Space Administration (CNSA) chief, Luan Enjie, affirmed his country's intention to conduct lunar exploration, although little in the way of detail was forthcoming thanks to China's notorious state-imposed secrecy. At the time, China had yet to place a man in orbit and Enjie's comments were largely ignored by the Western media. However, in October 2003, China's first astronaut, or *taikonaut*, People's Liberation Army Lt. Col. Yang Liwei was conferred with the title of "Space Hero" upon returning from the country's first manned spaceflight. The event prompted reporters in the West to ponder whether the next footsteps on the Moon might be Chinese. When China repeated its manned spaceflight feat two years later with the launch of Shenzhou 6 (Figure 2.1), carrying taikonauts Fèi Jùnlóng and Niè Hǎishèng on a five-day mission in LEO the rumors of a Chinese manned lunar mission gained even more credibility.

China's lunar history

China's motivation for placing its taikonauts on the lunar surface is fueled largely by political motives of international prestige and self-aggrandizement. The political motives notwithstanding, the country's scientific community argues lunar exploration will help acquire knowledge about the Moon and help prepare for future deep-space missions.

Preliminary studies for China's lunar exploration commenced in 1993 when a team of scientists led by Ouyang Ziyuan and Zhu Guibo of the China Aerospace Industry Corporation performed feasibility studies assessing the nation's space technology and infrastructures. The studies concluded in 1995, suggesting it would be possible to orbit a lunar satellite by 2000. Two years later, in April 1997, Chinese Academy of Sciences (CAS) members Yang Jiachi, Wang Daheng, and Chen Fangyun published the *Proposal for Development of Our Nation's Lunar Exploration Technology*, a document which led to preliminary research and development of robotic rovers for lunar exploration.

Lunar exploration gained more momentum in May 2000 and January 2001 when Tsinghua University organized two symposia on the subject of lunar exploration technology. A third lunar conference was held in March 2001 at Beijing's University of Aeronautics and Astronautics (BUAA), which discussed China's lunar exploration and human spaceflight programs and unveiled a feasibility study for sending Chinese astronauts to the lunar surface. The study suggested a five-step plan, the first of which would be a survey of the lunar surface via remote sensing to evaluate suitable soft-e landing sites. Although at this time there was much talk about lunar exploration,

stage which will be robotic exploration using rovers. The robotic exploration stage, which will occur between 2010 and 2020, will be followed up by sample return missions between 2020 and 2030, at which point China will be preparing to focus on human missions and the construction of a lunar base.

Despite Chinese officials stating that their lunar ambitions are a long time in the future, the aforementioned timeline represents what may be a conservative lunar schedule as a series of planned Shenzhou flights (Table 2.1) indicate China may be flying to the Moon sooner than 2030.

The launch schedule of Shenzhou spacecraft described in Table 2.1 indicates a logical, progressive, and methodical progression in goals. The goals seem to indicate China is pursuing ambitions targeted initially at establishing a space station before moving on to the main goal of landing its taikonauts on the Moon. In fact, on January 4th, 2003, Xu Yansong, a senior official of the CNSA, declared that "China will put men in space in the next six months and send a flyby mission to the Moon in four years." Yansong's statement was followed a month later by another declaration, this time by Huang Chunping, General Director for Launch Vehicles, who stated that "China has the full capability to send astronauts to the Moon." Another month passed and Yansong's and Chunping's statements were refuted when Ziyuan announced piloted Moon missions were not on the agenda for the next decade, a timeline that places the prestigious Shenzhou circumlunar mission in the 2013

Table 2.1. Planned Shenzhou missions.

Mission designation	*Date*	*Crew*	*Mission objective*
Shenzhou 7	October 2008	3	Extravehicular activity
Shenzhou 8	2009–2010	Unmanned	Launch of space laboratory module with docking ports
Shenzhou 9[a]	2009–2010	Unmanned	Rendezvous and docking with Shenzhou 8
Shenzhou 10	2010	2–3	Docking with Shenzhou 8 and Shenzhou 9
Shenzhou 11	2010	Unmanned	Launch of space laboratory module with docking ports
Shenzhou 12	2012	Unmanned	Rendezvous and dock with Shenzhou 11
Shenzhou 13	2012	2–3	Docking with Shenzhou 11 and Shenzhou 12

[a] This flight will carry China's first female taikonaut.

timeframe. Nevertheless, such a mission would be feasible today given the present capabilities of the Chinese space program.

Technology

As a result of their manned and lunar probe missions the Chinese now have a significant space infrastructure, comprising ground control and tracking facilities, launch centers, and vehicle assembly facilities. As a part of their tracking capability the Chinese are able to control satellites in geosynchronous orbit, a more complex operation than a simple circumlunar mission. Furthermore, their tracking ships are already operational for worldwide coverage for short-term missions such as one to the Moon.

Furthermore, China's Shenzhou spacecraft, used to carry Yang Liwei into orbit, is similar to Russia's Soyuz spacecraft. The Soyuz was flight-proven during the abortive Russian L1 manned lunar program of the 1960s and would be capable of re-entry simply by adding a thicker layer of ablative material to the capsule.

The two-launch architecture required for a circumlunar mission could be achieved using existing CZ-2E/CZ-2F boosters and fitting a docking system to an existing rocket stage. A manned Shenzhou could be launched by a CZ-2F Launch Vehicle (LV) followed by a restartable liquid oxygen (LOX)/liquid hydrogen (LH$_2$) stage launched by a CZ-2E LV. After the Shenzhou had made its first orbit, it would rendezvous and dock with the stage, which would then fire its engines, placing the combined Shenzhou and stage into a translunar trajectory.

Although landing taikonauts on the Moon is a complex operation, a manned circumlunar mission or a lunar orbital mission may be accomplished with much less hardware, complications, and expense. In fact, circumlunar missions are so achievable the popular space tourism company, Space Adventures Ltd., is selling circumlunar tickets at $100 million. Although Chinese statements are so vague they are open to all manner of interpretation, it is likely the Shenzhou 8, 9, and 10 missions will form a platform for evaluating hardware and procedures for a lunar mission that may take place shortly after. If the Chinese are successful in their lunar ambitions, they will deliver the most potent propaganda coup their space program can hope to achieve in its present condition.

JAPAN'S LUNAR INTENTIONS

In April 2005, Japan's Aerospace Exploration Agency (JAXA) announced an ambitious plan to send their astronauts to the Moon by 2025. The statement was made just two months after JAXA had witnessed its first successful launch since November 2003 when an unmanned H-2A LV malfunctioned shortly after launch and was destroyed mid-flight, an accident forcing the agency to ground the H-2A LV for 15 months. Until the H-2A failure JAXA had been pursuing a program focused primarily on unmanned probes, but following a government panel in 2004 the decision was taken to focus its attention on manned spaceflight instead. However, plans for its unmanned Kaguya mission remained intact, and were realized in September 2007 when the previously troublesome H-2A LV launched a three-tonne orbiter on its way to the Moon, where it released two small satellites into polar orbit. The launch represented the first stage in Japan's route to the Moon and although JAXA officials played down talk of a space race with its neighbor, China, the significance of the Kaguya mission as a step towards Japan's plans to establish a manned base on the Moon were hard to ignore.

Japan's lunar architecture

Next on Japan's lunar exploration schedule, according to Manabu Kato, chief scientist overseeing the Kaguya project, is to land a robot on the surface of the Moon in 2012, followed by a sample return mission in 2018. "We are also discussing human exploration but we expect international collaboration," Kato told reporters while attending a space conference in Hyderabad in September 2007. Kato added human exploration could be followed by human bases on the Moon and cooperation between nations for exploring the Moon should be based on the model currently implemented by nations collaborating on the International Space Station (ISS). Kato did not rule out collaboration with China or India but indicated much would need to be discussed before such cooperation were to take place.

RUSSIA'S MANNED MOON PROGRAM

During a new conference on January 11th, 2007, Nikolai Sevastyanov, head of the Russian spacecraft manufacturer Energia, said "Our motto is 'toward Mars with a stop on the Moon'." The comments by Sevastynanov do not reflect current Russian space policy, however.

Four years after President Bush's Vision for Space Exploration (VSE), the goals of the Russian Space Agency remain undefined. On April 11th, 2008, on the eve of Cosmonautics Day, Anatoly Perminov, the head of the Russian Space Agency, indicated there were plans for next-generation spacecraft and a new launch vehicle capable of lunar expeditions. He also reiterated the goal of establishing Russia's own space station for the purpose of assembling spacecraft destined for the Moon and Mars. To date, no timeline for these goals has been announced, although Russia has commenced development of new spacecraft that may enable lunar missions.

Soyuz K

According to RIA Novosti, the Russian News and Information Agency, Russia's Rocket and Space Corporation (RSC), Energia, will develop and build a new space-craft capable of flying to the Moon by 2012. The working name of the new spacecraft, *Soyuz K*, will be designed to be launched from either the Baikonur space center or the equatorial site at Kourou in French Guyana and will form the key element in Energia's three-stage plan to explore the Moon. According to Nikolai Sevastyanov, President of Energia, these stages of lunar exploration will involve manned Soyuz flights to the Moon followed by the construction of a permanent base, which would occur between 2010 and 2025, and finally the industrial exploration of lunar resources such as the extraction of helium-3.

Kliper

In addition to designing and developing Soyuz K, Energia is also working on creating a new Russian space shuttle, Kliper (Figure 2.4). Capable of carrying six cosmonauts, Kliper will be designed to replace the aging Soyuz and Progress craft and will eventually make flights to and from the ISS and also to the Moon and possibly Mars. However, on July 18th, 2006, the state tender of the Russian Space Agency for Kliper was canceled. Although development of Kliper continues, Energia may require private investment if the spacecraft is to be ready for its first test flights scheduled for 2012.

Figure 2.4. On display at the 2005 Paris Air Show, the Kliper is a six-person spaceplane intended to replace the Soyuz and Progress crew and cargo vehicles. Russia's Energiya Rocket and Space Corporation plan to have the vehicle operational by 2015. Image credit: Ian Murphy.

a lunar lander and two space tugs. The two tugs would be used to boost the lunar lander and the Soyuz with crew on a lunar trajectory. The Soyuz would be launched by a Soyuz 3 rocket. Such a mission profile would require four launches, two rendezvous and docking operations in LEO, and two similar operations in lunar orbit

Although the Russian manned Moon program sounds similar to NASA's VSE announced by President Bush in January, 2004, the Russians have an unquestioned commitment to manned spaceflight, possess excellent space technology, and have a vast reservoir of experience in the area of long-duration spaceflight.

TAKING EUROPE TO THE MOON

"We think it is technically feasible to have a manned mission to the Moon between 2020 and 2025 and then to Mars between 2030 and 2035. We need to go back to the Moon before we can go to Mars."

Dr. Franco Ongaro, Project Manager of ESA's Aurora Space Exploration Program

In 2003, the Human Spaceflight Vision Group (HSVG) was formed with the purpose of creating a vision for European human spaceflight up to 2025. Experts from eight countries representing the fields of communications, marketing, research, and academic institutes supported by the ESA assessed the needs and interests of various stakeholders with interests in human spaceflight and announced the following vision statement:

"In 2025, Europe will begin to operate a permanently manned outpost on the Moon as part of a multi-decade, international exploration effort to serve humanity, thus increasing our knowledge and helping us to address the global challenges of the future."

After much discussion and feasibility analyses ESA identified high-priority study objectives specific to lunar exploration. Among the objectives were items such as determining the number of launches required, calculation of delta-V requirements, determination of component assembly in LEO, definition of mission scenarios, and lunar base assembly strategies. This information was assessed as part of ESA's ESTEC Concurrent Design Facility (CDF) study, which formulated a mission architecture based on evolution of the Ariane 5 capability.

Concurrent Design Facility lunar mission

The CDF's manned mission to the Moon requires three modified Ariane 5 launches to assemble the Lunar Vehicle in LEO before the crew arrives onboard a Soyuz. The vehicle elements would rendezvous and dock in LEO before being sent into a lunar transfer trajectory. Initially, a number of demonstration missions will be flown before launching elements designed to establish an outpost on the Moon. Outpost elements would include a Habitation Module (HM), a Logistics Module (LM), and a Resource Module (RM).

Manned mission to the Moon

A mission architecture that also utilizes ESA's Ariane 5 (Figure 2.6) capability is the Manned Mission to the Moon (M^3), a project created as a result of a thesis by students at the ESA/ESTEC Education Office.

> "Put a European on the Moon by providing a spacecraft able to land on and safely return from the lunar South Pole. Prove that it is possible with the use of today's technology. Utilize Ariane 5 and decide on predecessor missions when required. Undertake lunar activities and retrieve hydrogen compound samples from a South Pole crater of the Moon. The mission will be a corporate effort of ESA and research institutes, opening up new scientific and educational opportunities."
>
> M^3 mission statement, given by Professor W. Ockels and E. Trottemant
> (November 2003)

The M^3 mission goals are to put a European on the Moon with crew safety as a primary objective and to use as much commercial off-the-shelf technology to achieve that goal. In common with NASA's Constellation Program, the M^3 program intends to plan and rehearse mission phases and to develop support systems in advance of any manned mission. The LV chosen for the undertaking is the Ariane ESC-B LV, which will be launched from Kourou, French Guyana.

Mission plan

The mission plan calls for a five-day outbound and five-day return lunar transfer and a surface stay of nearly two weeks for a total mission duration not exceeding 25 days. The preferred landing site is the lunar South Pole in the vicinity of the Shackleton Crater, chosen due to the availability of hydrogen compounds, illumination conditions that provide visibility of the Sun for 80% of the mission duration, and a maximum communication window with Earth. ESA's lunar astronauts will have a lunar vehicle to assist them in fulfilling science objectives such as collecting lunar samples and sample drilling for hydrogen compounds. On their return trip the crew will take advantage of an initial transfer trajectory providing a guaranteed free return to Earth, a journey that will be tracked live using the ESA ESTRACK ground station network.

Mission hardware

Originally, M^3's mission plan was to utilize two LVs, the primary LV being the Ariane 5 ECB, and the second being the Russian Soyuz TMA, carrying two

Figure 2.6. The European Space Agency's Ariane 5 Launch Vehicle could be used to send vehicle elements for a manned mission to the Moon. Image courtesy: ESA.

Table 2.3. M^3 Launch Vehicles.

Soyuz TMA		Ariane (ECA variant)[a]		
Crew	2–3	Height	59 m	
Orbit parameters	51.6°	Diameter	5.4 m	
Altitude of insertion	5.2°	Mass	777,000 kg	
Altitude during spacecraft docking	~425 km	Stages	2	
Mass of delivered payload	≥100 kg	Payload to LEO	21,000 kg	
Mass of returned payload	≥50 kg		First-stage Ariane 5 ECA	Second-stage Ariane 5 ECA
Flying life (including autonomous flight time)	200 days	Engines	1 Vulcain 2	1 HM7-B
Touchdown speed main parachute system	Max: 1.4–2.6 m/s	Thrust	1,340 kN	64.7 kN
Launch Vehicle	Soyuz FG	Specific impulse	431 s	446 s
		Burn time	650 s	960 s
		Fuel	LH$_2$/LOX	LH$_2$/LOX

[a] Similar to the ECB.

crewmembers into orbit. Unfortunately, the plans to use the Ariane 5 ECB, which had been scheduled to enter service in 2006, have been put on hold due to budget cuts. At an ESA conference in December 2005 no decision was made to restart or cancel the program, which means it is possible development may resume following the ESA conference in 2008.

The rationale for using the Ariane 5 ECB in combination with the Soyuz TMA was to avoid the problems associated with human-rating the Ariane LV. Also, using the Ariane 5 ECB avoided the problems of launching the crew from Kourou, French Guyana, the location of which provides restrictive launch windows for the purpose of launching to the ISS. However, since the M^3 mission plan will rely on the Ariane to launch mission elements, some of these elements will have to remain in LEO for a month or more due to the limited Ariane 5 launch frequency from Kourou.

M^3's lunar mission hardware (Table 2.3) consists of support and main mission modules. The support modules consist of elements that support lunar exploration and the main mission modules consist of elements required to transfer the crew from the Earth to the Moon and return them from the lunar surface back to Earth.

Table 2.4. M3 support missions [1].

Launch designation	Launch date	Module	Description
Launch Support 1	23-8-2014	Robotic Descent Module (RD)	• Launched into LTO • Carries robotic rover and satellite • Provides delta-V burn required to reach LLO
		Robotic Rover & Storage (SR)	• Confirms presence of hydrogen compounds in Shackleton Crater • Retrieves/returns samples to crater rim
		Satellite & Storage (SS)	• Used to investigate lunar surface from LLO • Provides information for landing site • Provides communication relay between mission vehicle and Earth ground stations
		Beacons & Storage (SB)	• Landing beacons dropped from orbit to aid automated landing
Launch Support 2	25-9-2016	Crew Descent Support (CDS)	• CDS module launched in LTO • Provides delta-V burn required to reach LLO • Docks with Crew Ascent Support module and Escape Habitat (EH) module in LLO
Launch Support 3	22-10-2016	Support Ascent Escape Habitat (SAEH)	• SAEH comprises ascent stage and EH • SAEH launched into LTO • Provides delta-V burn required to reach LLO • Docks with CDS module in LLO • CDS descends SAEH to lunar surface • SAEH provides back-up ascent vehicle if MAH fails

Mission architecture

The architecture comprises two distinct phases, based on an Ariane 5 capability of launching 27 tonnes to LEO. The first phase of support missions (Table 2.4) consist of three launches to place a suite of unmanned spacecraft in lunar orbit in preparation for the next phase comprising the launch of main mission modules (Table 2.5).

Under ESA's Aurora plan for space exploration a manned mission to the Moon is proposed for 2024 for the purpose of testing life support systems and ways to use *in*

Table 2.5. M3 main mission modules [1].

Launch designation	Launch date	Module	Description
Launch Mission 1	20-11-2016	Crew Descent Mission Kick Stage (CDMKS)	• CDMKS transfers CDM module from LEO to LTO • Launched in LEO where it docks with CDM • Provides delta-V to insert CDM to LTO and LLO
Launch Mission 2	17-12-2016	Crew Descent Mission (CDM)	• Launched to LEO and docks with CDMKS • CDMKS inserts CDM module to LTO • In LTO, CDMKS provides delta-V burn required to reach LLO • In LLO, CDMKS discarded • CDM docks with MAH and descends to lunar surface
Launch Mission 3	13-1-2017	Mission Ascent Habitat (MAH)	• MAH is combined ascent stage and habitat • MAH launched into LTO • Provides delta-V burn to reach LLO • In LLO, MAH docks with CDM module • CDM descends MAH to lunar surface • At lunar surface, MAH provides crew with ascent vehicle to reach LLO • Unpressurized rover and storage (URS) • Extends mobility of astronauts during surface sorties • Able to carry astronauts up to 15 km • Descends to lunar surface in CDM
Launch Mission 4	19-2-2017	Service Module (SM)	• Launched into LEO with CRC • SM-CRC docks with Soyuz in LEO to transfer crew to CRC • Provides delta-V burn required to reach LTO and LLO • SM-CRC docks with CDM-MAH in LLO. When crew return from lunar surface they will dock with SM and transfer to CRC. SM provides delta-V for transit from LLO to Earth
		Crew Re-Entry Capsule (CRC)	• Serves as habitat during transfer from LEO to LLO and during transfer to Earth • Launched in LEO together with SM

3

Next-generation launch vehicles

What you're looking at here is America's drive to explore the Moon, a space program coming back to life, a re-awakening, a renaissance, a re-dedication of our nation that will lead the way into the next chapter of human destiny.

Legendary NASA Flight Director, Gene Kranz
addressing the Constellation Program workforce, April 2008

OVERVIEW OF THE LAUNCH VEHICLE SELECTION PROCESS

The requirements of the Constellation Program mean that NASA must transition from being an agency focused on low Earth orbit (LEO) operations to developing transportation modes capable of exploring the Moon, Mars, and beyond. As the Space Shuttle is capable of flying only to LEO, and because of the vehicle's complexity and high fixed operating costs, NASA chose to retire the Shuttle by September 2010. To support manned lunar missions a new family of launch vehicles (LVs) was required and the Exploration Systems Architecture Study (ESAS) team was chartered to develop and evaluate viable launch system configurations for a Crew Launch Vehicle (CLV) and a Cargo Launch Vehicle (CaLV).

Selecting the Crew Launch Vehicle

As a part of the process of choosing a CLV and CaLV, the ESAS team developed candidate LV concepts derived from elements of the existing Evolved Expendable Launch Vehicle (EELV) fleet, and the Space Shuttle system. The ESAS team then compared each concept against the other for cost, reliability, safety, versatility, and extensibility (Table 3.1 and Figure 3.1).

Table 3.1. Original ESAS comparison of crew launch systems to LEO [5].

	Human-rated Atlas V (new US)	Human-rated Delta IV (new US)	Atlas Phase 2 (5.4 m core)	Atlas Phase X (8 m core)	Four-segment RSRB (one SSME)	Five-segment RSRB (with J-2S)	Five-segment RSRB (with four LR-85)
Payload 28.5°	30 t	28 t	26 t	70 t	25 t	26 t	27 t
Payload 51.6°	27 t	23 t	25 t	67 t	23 t	24 t	25 t
DDT&E^a	1.18	1.03	1.73	2.36	1.00	1.3	1.39
Average cost/flight^b	1.00	1.11	1.32	1.71	1.0	0.96	0.96
LOM (mean)	1 in 149	1 in 172	1 in 134	1 in 79	1 in 460	1 in 433	1 in 182
LOC (mean)	1 in 957	1 in 1,100	1 in 939	1 in 614	1 in 2,021	1 in 1,918	1 in 429

LOM = loss of mission; LOC = loss of crew; US = Upper Stage; RSRB = Reusable Solid Rocket Booster; SSME = Space Shuttle Main Engine.
^{a} DDT&E = design, development, test, and evaluation.
^{b} Based on an average of six launches per year.

After several trade studies, NASA finally opted for a two-stage, series burn CLV concept derived from elements of the existing Space Shuttle system that positioned the CEV on the nose of the vehicle. The first stage (FS) is derived from the Space Shuttle's Reusable Solid Rocket Motor (RSRM) and was originally composed of four field-assembled segments. This design was later changed to a five-segment

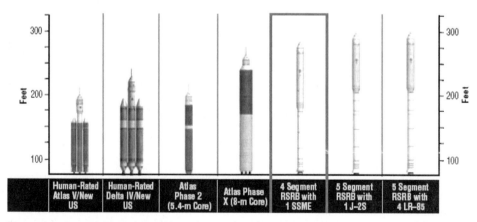

Figure 3.1. Launch Vehicle configurations evaluated by the Exploration Systems Architecture Study. Image courtesy: NASA.

version when it became apparent the vehicle would not be capable of generating the necessary lift to launch the Crew Exploration Vehicle (CEV) into LEO.

Selecting the Cargo Launch Vehicle

The ESAS team conducted a similar series of trade studies to determine the choice of the CaLV, considering EELV-derived and Shuttle-derived options, as summarized in Table 3.2.

The ESAS team's preferred CaLV configuration (Figure 3.2) is a stage-and-a-half vehicle comprised of two five-segment RSRMs and a large liquid oxygen (LOX)/ liquid hydrogen (LH$_2$) core vehicle that will carry 38% more propellant than the current Shuttle External Tank (ET).

Ares I and Ares V

The CLV, christened *Ares I* (Figure 3.3), will be the LV that will carry the CEV, comprising a Crew Module (CM) and Service Module (SM), into orbit by 2015. The unmanned CaLV, christened *Ares V* (Figure 3.4), will carry an Earth Departure Stage (EDS) and the Lunar Surface Access Module (LSAM), called *Altair*, by 2020. *Ares*, the ancient Greek name for the Red Planet, was chosen to reflect the choice of Mars as one of the intended destinations of the Vision for Space Exploration (VSE) while

Table 3.2. Original ESAS Comparison of Cargo Launch Systems to LEO [5].

	Five-segment RSRB in-line (five SSME)	Atlas Phase X 8 m core	Atlas Phase 3A 5.4 m core	Five-segment RSRB in-line (four SSME) core	Four-segment RSRB in-line (three SSME) core	Five-segment RSRB sidemount (three SSME)	Four-segment RSRB sidemount (three SSME)
Payload 28.5°	106 t (125 t with U.S.)	95 t	94 t	97 t	74 t	80 t	67 t
Lunar LV DDT&E	1.00	1.29	0.59	0.96	0.73	0.80	0.75
Lunar LV average cost/ flight	1.00	1.08	1.19	0.87	0.78	1.13	1.13
LOM Cargo (mean)	1 in 124	1 in 71	1 in 88	1 in 133	1 in 176	1 in 172	1 in 173
LOC (mean)	1 in 2,021	1 in 536	1 in 612	1 in 915	1 in 1,170	N/A	N/A

Figure 3.2. The Cargo Launch Vehicle, Ares V, is a Shuttle-derived Launch Vehicle consisting of two five-segment Reusable Solid Rocket Boosters and an External Tank. Image courtesy: NASA.

the *I* and *V* designations pay homage to Apollo's Saturn I and Saturn V that enabled NASA astronauts to land on the Moon in 1969.

NASA's rationale for choosing Shuttle-derived launch vehicles

NASA's decision to move forward with the Ares family of LVs represented the outcome of careful consideration, study, and evaluation of hundreds of commercial, government, and concept LV alternatives and architecture systems that could be used for human space exploration. When NASA performed these evaluations, the agency considered factors such as desired lift capacity, reliability, and the life cycle development costs of different approaches. The culmination of these efforts was the release in 2005 of the ESAS [5], which analyzed architecture and LV requirements to provide the safest, most reliable, and cost-effective system architecture for returning humans to the Moon. Following exhaustive consideration and examination of EELV-derived, Shuttle-derived, and "clean sheet" architectures, it was determined that, based on human safety, programmatic risk, reliability, and mission performance, the architecture that offered the most advantages was one that relied on Shuttle-derived technology. The decision was supported by evaluation of numerous current flight and test hardware databases, a factor that provided an added safety margin when it came to defining the architecture's technology and safety. Some of the decisions that resulted in the choice of a Shuttle-derived architecture are described here.

Figure 3.3. Ares I.
Image courtesy:
NASA.

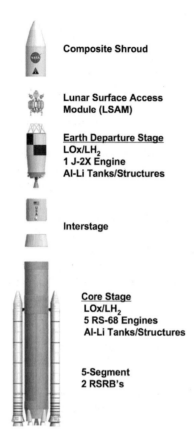

Composite Shroud

Lunar Surface Access
Module (LSAM)

Earth Departure Stage
LOx/LH$_2$
1 J-2X Engine
Al-Li Tanks/Structures

Interstage

Core Stage
LOx/LH$_2$
5 RS-68 Engines
Al-Li Tanks/Structures

5-Segment
2 RSRB's

Figure 3.4. An exploded view of Ares V. Image courtesy: NASA.

ESAS considerations

A major consideration faced by the ESAS study group was the issue of vehicle performance. One crew transport option on the table was to use an EELV from either the currently used heavy-lift Delta IV or the retired Atlas V family. However, an evaluation of the medium-class EELVs revealed that they underperformed by 40% to 60%. Also, an assessment of the heavy EELV class revealed that, although the lift capacity was closer to what was required to launch the 23.3-tonne CEV, major modifications in avionics, telemetry, and engines would be needed to human-rate the vehicles. In addition to the costs of these modifications an additional cost would be incurred if either the Atlas or Delta heavy classes were selected since both would require the development of new upper stages to achieve the necessary lift performance.

Another vehicle performance consideration revealed by the ESAS assessment was the need to minimize the on-orbit assembly complexity required by lunar missions. To achieve this goal the ESAS imposed a limit of no more than three launches for a lunar mission, which resulted in a lunar architecture requiring an LV capable of launching nearly a hundred tonnes into LEO. If the decision had been made to use existing EELVs a typical lunar mission would have required as many as seven launches, a number that would have exponentially increased the risk of loss of mission (LOM) and/or loss of crew (LOC). Although it was determined that current EELVs could have been developed to produce a 100-tonne LEO LV the decrease in mission safety (Table 3.2) as a result of the necessary modifications (such as two upper stages and adding strap-on core boosters) to achieve such a goal was determined to be excessive.

Another safety aspect driving the decision of whether or not to choose a current EELV was the issue of crew safety and reliability, since the aforementioned EELV families were originally designed to carry unmanned payloads. Choosing to modify an existing EELV therefore would not only have required extensive changes to flight termination software, abort scenarios, and inflight crew control capabilities, but would also have required NASA to modify launch pads by building emergency egress infrastructure.

Another assessment made by the ESAS group included an evaluation of risk to crew safety based on the human rating and reliability of systems such as propulsion and boosters. Based on this analysis it was deemed that the Shuttle, thanks to its extensive history (the Challenger and Columbia accidents notwithstanding) was the safest vehicle compared with the EELV families. Although it may seem surprising that a vehicle in which 14 astronauts have died was judged safe, it must be remembered that the comparison was with vehicles designed for unmanned payloads. The Shuttle, by virtue of being qualified to carry humans, is a vehicle in which each system is already human-rated and in which each structure complies with NASA's required 1.4 Factor of Safety (NASA-STD-5001), whereas the current families of EELVs only have a structural Factor of Safety (SF) of ≤ 1.25. Obviously, if an EELV had been chosen, each structure and system would have required a lengthy modification process involving costly redundancy upgrades, an effort that would have necessitated a significant engineering and development program.

Another major consideration the ESAS team had to wrestle with was the issue of life cycle cost. Unsurprisingly, given the design, development, testing, and evaluation (DDT&E) requirements to upgrade existing EELVs to human-rated status, it was determined that a Shuttle-derived LV combination, utilizing hardware and systems already human-rated, would incur significantly lower non-recurring costs. Although many observers perceive that given the decision to use a Shuttle-derived architecture, the EELV infrastructure and capability would be lost, this will not be the case since NASA intends to use existing EELVs for robotic exploration missions.

Despite occasional adverse publicity directed at NASA's LV selection and the ESAS team's recommendations, several external reviews have been conducted supporting the choice of a Shuttle-derived Ares I and Ares V. In 2005, the Department of Defence (DoD) reviewed NASA's analysis and agreed with the agency's approach, a sentiment validated in October 2006, by the Congressional Budget Office (CBO), which concluded that the mission reliability, mission mode, predicted LV safety and reliability, and budget projections put forward by NASA were sound. The most recent report by the Government Accountability Office (GAO), published in November 2007, noted that:

"The Ares I project was also proactive in ensuring that the ongoing project was in compliance with NASA's new directives, which include elements of a knowledge-based approach. NASA's new acquisition directives require a series of key reviews and decision points between each life cycle phase of the Ares I project that serve as gates through which the project must pass before moving forward. We found that the Ares I project had implemented the use of key decision points and adopted the recommended entrance and exit criteria for the December 2006 Systems Requirements Review and the upcoming October 2007 Systems Definition Review."

Designing a transportation architecture that will be sufficiently versatile and capable of performing immediate and future exploration requirements demanded that NASA consider a myriad of factors. These factors ranged from maximizing the use of existing human-rated systems and infrastructure to creating the most direct growth path for the exploration requirements of missions to the Moon, Mars, and beyond. In deciding on the Ares I and Ares V architecture NASA has ensured it is in a position to meet the goals and objectives of the exploration mission.

DESIGNING ARES

The technical management that oversees the development and design of NASA's new LV family is the responsibility of the Exploration Launch Projects Office (ELPO), located at Johnson Space Center (JSC). The ELPO was chartered by the Constellation Program to deliver safe and reliable crew and cargo LVs designed to minimize life cycle costs and to meet the operational requirements of the spectrum of missions mandated by the VSE.

Applying a "test as you fly" philosophy, the ELPO's project engineers utilize a variety of applications ranging from computer-aided modeling and simulation procedures to subscale wind tunnel models and real-world testing to not only yield information on which to base critical hardware decisions but also to gain confidence in the systems being built. The ELPO's approach of utilizing rigorous systems, engineering, and standards is guided by a framework of internal and external independent milestone reviews (Table 3.3) that define specific success criteria at each level of LV design and development.

ARES I

Design history overview

The evolution of Ares I can be traced back to 1995 when Lockheed Martin submitted a report of work conducted under an Advanced Transportation System Studies (ATSS) contract [1] to Marshall Space Flight Center (MSFC). The summary of the ATSS report described LV configurations similar to the current Ares I design, featuring liquid rocket second stages stacked above SRB first stages and variants of these designs that included the possible use of the J-2S engine for the upper stage (US).

In January 2004, nearly nine years after the ATSS report, President Bush

Table 3.3. Project Technical Reviews [5].

Review	Acronym	Purpose
System Requirements Review	SRR	• Verifies requirements are correctly defined, implemented, and traceable • Assures hardware and software is designed and built to designated configuration
Preliminary Design Review	PDR	• Provides completed design specifications, life cycle estimates and identifies long-lead items • Provides manufacturing plans and verifies design specifications are baselined • Confirms design is 30% complete
Critical Design Review	CDR	• Discloses complete design and determines that any technical anomalies have been resolved • Verifies design maturity and justifies decision to commence manufacturing • Confirms design is 90% complete
Design Certification Review	DCR	• Operates as a quality assurance mechanism • Ensures system(s) can achieve its mission
Flight Readiness Review	FRR	• Examines, tests, analyzes, and audits system's readiness for flight after system has been configured for launch • Project Manager and Chief Engineer certify system ready for flight

announced the VSE, which resulted in NASA chartering the ESAS to determine the requirements and configurations for crew and cargo systems for exploration missions to the Moon and Mars. This task resulted in the selection of a Shuttle-derived launch architecture which initially envisaged using a standard four-segment SRB for the FS and a SSME variant for the US. However, although the initial design was approved, it was quickly determined that the Orion CEV would be too heavy for the planned four-segment LV, a finding that spurred NASA to reduce the size of Orion and add an extra segment to the SRB FS. Also, rather than use the single SSME as suggested in the original design, the revised option chose to use the Apollo-derived J-2X engine. This choice was due in part to the latter engine's advantage of being able to start in mid-flight and in a vacuum, whereas the SSME was designed to start only on the ground.

Design endorsement

In October 2006, NASA announced it would extend a previous contract with ATK Thiokol of Brigham City, Utah, to continue design and development of the Ares I FS.

Under the contract ATK Thiokol was required to maintain a DDT&E schedule for the FS, to expedite procurement of new nozzle hardware and maintain design and engineering analysis that would lead to an SRR. The SRR, completed on January 4th, 2007, confirmed Ares I system requirements had been validated and fulfilled mission requirements and endorsed the Ares I architecture design concept. The following day NASA extended the Ares I development contract with ATK Thiokol. Under this contract ATK Thiokol was tasked with maintaining design and engineering analysis, to prepare pilot parachute development tests and to support the initial test flight scheduled for the spring of 2009.

The announcement of the SRR, endorsement of the Ares I design, and the contracts with ATK Thiokol were the culmination of a lengthy process of LV design evolution which had resulted in Ares I undergoing a number of configurations, some of which predated the ESAS.

Ares I design evolution

Figure 3.5 shows how Ares I evolved from the pre-ESAS SRB launcher variant to the present design configuration. The first concept, which featured a standard four-segment SRB and a J-2S US, began in the NASA Astronaut Office and was later seen in releases from ATK and Boeing. It was designed to launch a small 18-tonne vehicle,

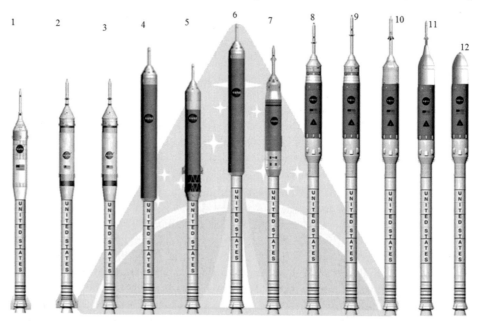

Figure 3.5. Evolution of Ares V concepts. Image courtesy: NASA. 1, pre-ESAS; 2, pre-ESAS ATK; 3, pre-ESAS ATK; 4, ESAS 4-segment + 1 SSME; 5, 4-segment + mesh interstage; 6, post-ESAS 5-segment + 1 J-2X; 7, CEV changes to 5.0 m diameter; 8, Orion 604 with updated LES; 9, Orion 605, tower height change; 10, Orion 606, shroud covers SM; 11, Orion 607; 12, Orion MLAS.

similar to the Apollo vehicle, to the ISS. The next pre-ESAS launch concept, which also used a four-segment SRB and J-2S US, first appeared on the *safesimplesoon.com* website and shows the US of a lengthened SRB's launcher. This was followed by a similar SRB launcher, this time with the fins removed (a variant that was the last pre-ESAS concept to be considered).

The first ESAS Ares I featured a four-stage SRB FS and an SSME US that grew considerably during the ESAS. Following the ESAS it was determined that without an SSME air start capability it would be necessary to use the J-2X, a change that resulted in an Ares I variant that utilized a five-segment SRB. This first post-ESAS Ares I was also the tallest one proposed. As the design of the various flight elements progressed, it became apparent a reduction in weight was required. To achieve some weight savings engineers reduced Orion's diameter to five meters, resulting in a shortening of the US due to the inclusion of a common bulkhead design and a shorter fairing adapter for the reduced size of Orion. This iteration was followed by the Orion 604 version of Ares I which featured updates to the baseline LES in addition to a protective boost cover for the CEV CM. A longer FS was the result of including an equipment interstage that contained the parachutes. Few changes are noticeable in the Orion 605 Ares I except for a longer LES tower. Similarly, the Orion 606 Ares I featured only a minor change to the SM and the encapsulation of the SM. The image of Orion 607 Ares I shows the optimized LAS design version whereas the Orion MLAS, which represents the final iteration of the Ares I, shows only a change to the MLAS LES, a modification that resulted in an overall height reduction.

Role of the Constellation Architecture Requirements Document

The design requirements of all the hardware, software, facilities, and personnel required to perform the Design Reference Missions (DRMs) is defined in NASA's 593-page *Constellation Architecture Requirements Document* (CARD). CARD is structured to provide mission planners and systems engineers with top-level design guidance and an overview of the architecture's functional and performance requirements. Examples of some of the CARD requirements for the CLV are listed in Table 3.4.

Ares I first-stage design and development progress

Design

The Ares I FS is a five-segment RSRB which will burn a specially formulated and shaped solid propellant called polybutadiene acrylonitrite (PBAN). Above the FS (Figure 3.6) sits a forward adapter/interstage designed to interface with the Ares I liquid-fueled US. Above the interstage is a forward skirt extension which houses the Main Parachute Support System (MPSS) and main parachutes for FS recovery. The frustrum, located at the top of the FS elements, provides the physical transition from the smaller diameter of the FS and the larger diameter of the US. Other modifications made to the original Shuttle SRB included removing the ET attachment points and altering the propellant grain inside the SRB.

Table 3.4. CARD Crew Launch Vehicle description and requirements [4].

CLV description
The CLV is the launch vehicle for the CEV. It consists of a five-segment solid rocket booster first stage and a cryogenic liquid hydrogen/oxygen fueled upper stage consisting of a structural tank assembly and a J-2X engine. The first stage is reusable and the upper stage is discarded after the CEV has separated during ascent.

CLV requirements

[CA5916-PO]   The CLV shall be single fault tolerant for critical hazards that do not cause abort or loss of mission.
Rationale: Single-fault tolerance provides for mission-critical failures and is dictated by programmatic decisions to ensure mission success. The Constellation Program will define levels of fault tolerance that are satisfied by multiple systems and the allocations to those systems. This does not preclude more than the minimum level of fault tolerance.

[CA1065-PO]   The CLV shall limit its contribution to the risk of loss of mission (LOM) for any mission to no greater than 1 in 500.
Rationale: The 1 in 500 means a 0.002 (or 0.2%) probability of loss of CLV mission for any Constellation DRM. This requirement is driven by CxP 70003-ANX01, Constellation Need, Goals, Objectives, Safety Goal CxP-G02: Provide a substantial increase in safety, crew survival, and reliability of the overall system over legacy systems.

[CA0389-PO]   The CLV shall use a single five-segment solid rocket booster modified from the Space Shuttle Solid Rocket Booster (SSRB) for first-stage propulsion and a single modified Apollo J-2X engine for second-stage propulsion.
Rationale: The CLV will take advantage of the flight-proven propulsion systems components developed for the Space Shuttle and Apollo. These launch vehicle components, which have supported over 100 Space Shuttle and numerous Apollo missions, have extensive test/flight experience databases available for CLV designers to leverage. In addition, CLV designers will be able to leverage the ground processing/production facilities, workforce, and tooling already in place to support Space Shuttle operations.

Development

Fabrication of the Ares I FS fifth-segment simulator and forward skirt began in November 2007 and are on schedule to be delivered to KSC in August 2008 to be ready for the Ares I-X launch on April 15th, 2009. Also complete is the pyro shock testing program which characterized shock loads on the avionics and reaction control thrusters.

Recent FS developments include a milestone Major Design Review (MDR), held in Ogden, Utah, on January 14–17, 2008, attended by representatives of MSFC Engineering, MSFC Safety and Mission Assurance, Ares I-X Mission Management Office (MMO), KSC Ground Operations, and KSC Ground Systems. The purpose of

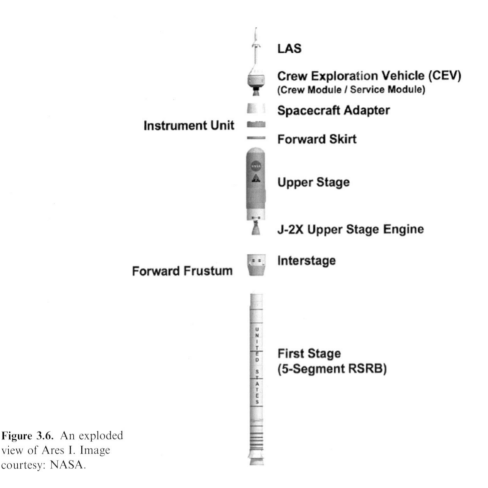

LAS

Crew Exploration Vehicle (CEV)
(Crew Module / Service Module)

Spacecraft Adapter

Instrument Unit

Forward Skirt

Upper Stage

J-2X Upper Stage Engine

Forward Frustum   **Interstage**

First Stage
(5-Segment RSRB)

Figure 3.6. An exploded
view of Ares I. Image
courtesy: NASA.

the review was to confirm the integrated FS design met baseline-stage requirements and that the FS was ready to support the Ares I-X CDR. Although the MDR generated approximately two hundred Requests for Action (RFAs), none was deemed to have any significant impact upon schedules or to present any major technical challenges.

Two months later the Ares I DSS MDR concluded with the MDR Board reviewing and screening action items pertaining to the DSS. The outcome of the DSS MDR was the granting of approval by the Board to start fabrication on the Ares I-X parachutes, a decision that culminated in the main parachute cluster drop test (CDT) at the Yuma Proving Grounds (YPG). These tests, conducted on February 9th, 2008, evaluated the parachute recovery system for the Ares I FS booster and involved a full-scale prototype pilot parachute being deployed at 5,500 m. Unfortunately, during one of the three tests, one of the parachute risers failed and the test was unsuccessful, although the test parachute was in usable condition following the test. The two other tests were deemed successful.

Ares I upper-stage design and development progress

Design

The Ares I US is being designed by MSFC and is based on the internal structure of the Space Shuttle's ET. The US was originally designed to incorporate separate fuel and oxidizer tanks separated by an intertank (a structure used in the Apollo program) but, due to mass restrictions, engineers decided instead to use a common bulkhead between the tanks.

The upper section of the US includes a spacecraft adapter (SCA) system designed to mate with the Orion CEV, while the lower section includes a thruster system to provide roll control for the FS and US. Power for the US will be provided by the LH_2 and LOX fueled J-2X rocket engine, described later.

Development

Recent developments in the progress and construction of the US include testing of the umbilical plates concept (UPC) and evaluation of quick disconnect (QD) compliance capabilities, which were performed at KSC in January 2008. The purpose of the tests was to determine the functionality of electrical connectors and pneumatic disconnects to ensure that when Ares I launches there is sufficient clearance between the vehicle and the plates.

Ares I avionics

The final major contract award for Ares I was made to the Boeing Company of Huntsville, Alabama, in December 2007, for the production, delivery, and installation of the Instrument Unit Avionics (IUA), which will provide guidance, navigation, and control for the entire LV. The IUA will include subsystems such as the J-2X engine interface, US Reaction Control System (RCS), FS Roll Control System, Hydraulic Power Unit Controller (HPUC), Data Acquisition/Recorder Unit, and the Ignition/Separation Unit. Power for all the subsystems will be provided by the Electrical Power System (EPS) that comprises batteries, power distribution and control units, DC-to-AC Inverter Units, and cabling. Located in the IUA, the EPS will ensure redundant sources of 28-volt direct current (VDC) from the time ground power is removed prior to launch until the end of the mission. In addition to the primary EPS, the US also features an independent EPS, located in the Interstage (Figure 3.7), the purpose of which is to provide power to the First Stage Roll Control System (FSRCS) thrusters.

Additional avionic subsystems such as the Developmental Flight Instrumentation (DFI) subsystem will be installed during the test flights. The DFI, mounted in the IUA, comprises data acquisition units (DAUs), a battery module unit (BMU), a telemetry system, and various sensors and will serve as a stand-alone element, whose purpose is to acquire and store data during the test flights, enabling engineers to analyze the performance of Ares I.

The Avionics Integrated Product Team (IPT), charged with the development of the IUA, held a successful CDR on November 27–29, 2007, an event that resulted in

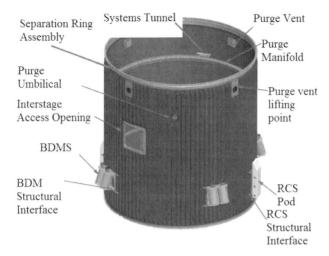

Figure 3.7. Ares I Interstage. Image courtesy: NASA.

fewer than 100 RFAs. Held concurrently with the CDR was an MRR for the First Stage Avionics Module (FSAM), which will house most of the avionics boxes located on the FS. As a result of successfully completing the FSAM MRR the Ares I Control Board granted approval for structural manufacturing to begin in January 2008.

Ares I safety systems

To prevent combustible LOX and LH$_2$ accumulating to dangerous levels while Ares I is on the launch pad the US is fitted with a Purge System. The Purge System ensures thermally conditioned inert nitrogen gas is pumped into the closed compartments of the US while at the same time exhausting excess nitrogen through special vents at the bottom of the compartments. Another safety system used on the launch pad is the Hazardous Gas Detection System (HGDS), which samples, detects, and measures the concentration of hazardous gases in the compartments of the US prior to launch.

Ares I test flights

When operational, Ares I will utilize a five-segment booster, but during the first test flight a four-segment booster will be used as the five-segment variant will not be ready. Instead, NASA will mount a fifth dummy segment on top of the standard four-segment version to replicate the correct aerodynamic, mass, and CG properties of the final design. Positioned above the booster will be the pilot, drogue, and main parachutes, which will be protected by an aero-shell cover, which will shield the parachutes from damaging blast following ignition of the US engine.

Nominal mission profile

When operational, a typical mission will commence with the Ares I FS booster lifting Orion to an altitude of 50 km at which point the FS will separate and fall back to the ocean to be recovered in the same manner as Shuttle SRBs are recovered today. The Ares I US will then take over and carry Orion to an elliptical orbit of 245 km. Once the US has separated, Orion's propulsion system will power the spacecraft to its 300 km circular orbit. There, the Crew Module (CM) and the Service Module (SM) will either rendezvous and dock with the ISS or continue to the Moon. Before this flight occurs, however, a series of test flights must be conducted.

First test flight: Ares I-X

NASA has learned through experience that operating costs can be reduced by thorough and incremental testing of its LVs. This "test as you fly" philosophy was implemented during the Saturn development program, which conducted multiple demonstration and verification flight tests before certifying the vehicle safe for humans.

A different tactic was adopted for the Space Shuttle program which involved much less testing but did not appreciably reduce schedule costs, which is why NASA plans to revert to the Saturn approach and conduct a progressive series of demonstration, verification, and mission flight tests before certifying Ares I safe for humans. The first of these flights will take place on April 15th, 2009, with the launch of the Ares I Flight Test Vehicle (FTV) on a suborbital development flight test.

The Ares I FTV, which will be integrated at KSC shortly before launch, will be very similar in mass and configuration to the operational vehicle and will incorporate both flight and mock-up hardware specific to the objectives of the test flight, although it will not have the capability to receive commands from the ground while in flight. In fact, the only active element of the Ares I FTV will be the four SRB segments, since the fifth segment will be empty of propellant.

Ares I-X flight profile

After lift-off from KSC, the Ares I FTV will climb to an altitude of 250,000 feet at which point it will be traveling at Mach 4.5 and be subjected to a maximum pressure quotient (Max Q) of nearly 800 pounds per square foot. At 132 seconds into the flight, FS burnout and US separation will occur, at which point the US simulator and the Orion crew vehicle and launch abort system simulator will separate from the FS. The simulator hardware will then fall into the Atlantic and will not be retrieved, whereas the FS booster will "fly" a complete recovery sequence, upon which the hardware will be retrieved and inspected at KSC.

NASA hopes the Ares I-X flight will fulfill a number of objectives such as demonstrating the Ascent Flight Control System, testing the FS Parachute Recovery System, validating assembly and processing flow, and characterizing the roll torque due to the FS motor performance.

The second test flight, designated Ares I-Y, is scheduled for 2012, and will demonstrate flight control algorithms with the five-segment SRB and a high-fidelity US simulator. It will also demonstrate a Launch Abort System (LAS) at high altitude, measure and characterize launch and ascent environments for the five-segment SRB, and demonstrate FS separation and recovery dynamics and performance. If all goes well, an unmanned flight designated Orion 1 will be launched in early 2013, which will eventually lead to the first human crew being transported to the ISS in 2015.

DEVELOPMENT PROBLEMS

"I hope no one was so ill-informed as to believe that we would be able to develop a system to replace the shuttle without facing any challenges in doing so. NASA has an excellent track record of resolving technical challenges. We're confident we'll solve this one as well."

NASA Administrator, Dr. Michael Griffen, Associated Press, January 20, 2008

The development of Ares I can often be measured by the progress of numerous Technical Interface Meetings (TIMs) that occur in the various Constellation Project Offices. Following a Project Office team meeting, recommendations are made and technical agreement is reached on issues ranging from the mechanical intricacies of passive stabilization damping for KSC ground connections to the esoteric complexities associated with subscale diffuser verification testing. These agreements go forward to the PDRs and are ultimately accepted or rejected, which, in the latter event, results in the issue being subjected to independent review or, if the matter concerns the performance of a system or hardware item, additional testing. Some of the more vexing problems concerning the design of the Ares I are described here.

Thrust oscillation

Perhaps the most troublesome performance matter that has concerned engineers is the issue of *thrust oscillation* (TO), a phenomenon also known as *resonant burning*. TO is a solid rocket motor (SRM) occurrence, characterized by increased acceleration pulses that may occur in the latter part of the FS phase of the flight. The pulses, which may vary in amplitude to the point at which structural damage is a possibility, are generated as a result of vortices created by burning propellant inside the SRM. The vortices may, as a result of flow disturbances, coincide with acoustic modes of the combustion chamber, thereby generating longitudinal forces which may in turn increase loads encountered by Ares I during ascent to orbit. According to NASA's computer-modeled performance of the Ares I FS, noticeable TO would occur at lift-off and would become progressively more severe, eventually peaking at 110 seconds into flight. If a certain frequency and amplitude of TO were to set up a specific resonance, worst-case vibration loads of between $4.5g$ and $5.5g$ would be experienced by the crew. Such a frequency would harm not only critical components such as the guidance, navigation, and control avionics but also completely incapacitate the crew.

Similar TO effects were noticed during the Shuttle SRB testing and during some Shuttle flights but there is insufficient information (DFI was only fitted for Shuttle flights STS-1 through STS-4) for the Ares I "Tiger Team" to predict exactly how the oscillation predicted for the Ares I FS might affect the LV and its crew. The Constellation Program's human system integration requirements set the Orion crew health vibration limit at less than $0.6g$ r.m.s. (root mean squared) in any axis over one minute during the ascent [5], stating that:

> "Vibrations beyond $0.6g$ r.m.s. for one minute are considered intolerable to humans. It is expected that internal organs could be damaged if the level of vibration or the time period for these levels were increased. In studies, subjects exposed to such levels for 1 and 3 minutes reported that they had to exert great effort to finish the test. Pain was reported primarily in the thorax, abdomen, and skeletal musculature. Varying effects on blood pressure and respiratory rate were also observed" [5].

Although the TO problem initially appeared to be a potentially destructive event, after a month of analysing the problem the Ares I Tiger Team suggested that TO may not be as severe as initially reported by space media analysts. In an attempt to resolve the problem Ares project engineers will add a suite of 750 instruments to the Ares I-X FTV hoping that pressure sensors, accelerometers, transducers, and other instruments will provide them with real data about vibrations from the vehicle's solid-fuel FS. The outcome of the data may result in expensive modifications having to be made to the Ares I and the Orion vehicle. These modifications may include fixes such as isolating specific components at risk of damage from oscillation and, possibly, to use isolation pads which, although adding weight to the Ares I, may, together with the other fixes, resolve the problem. For example, one mitigation package suggestion involves mounting the FS 14,000 lb recovery parachute package on springs to dampen the vibrations generated as propellant burns out, in addition to placing shock absorbers on the seats of the Orion vehicle. It is predicted that such a mitigation package would reduce the vibrations in the capsule to about $2g$, although this figure is still in excess of the $0.25g$ threshold required to guarantee the crew is not harmed and to ensure the astronauts can monitor displays and operate controls. Fortunately for NASA, given the number of different fixes available to engineers to solve the TO issue, there seems to be no evidence that the issue cannot be addressed within the planned budget and within the planned schedule.

Ares I critics and supporters

One of the main criticisms made of Ares I is that the production of an LV in the 25-tonnes payload class is not necessary given that similar LVs such as the Delta IV-Heavy already exist. Critics argue that not only would it have been more economical to have utilized an existing LV but, thanks to a proven track record, the safety margins would also be greater, although NASA counters this latter argument by rating the Ares I twice as safe as the Delta IV-Heavy.

Another criticism focuses upon the choice of two derivative development programs, one being the new five-segment SRB for the Ares I FS and its $3 billion development cost, and the other being the new J-2X rocket engine for the Ares I US and its $1.2 billion development cost. Critics argue that the development schedule with its inevitable slips, increased risk of unproven flight technology coupled with the extra cost cancel out the advantage of using Shuttle-derived hardware.

Another issue that has troubled space analysts and aerospace engineers is the technical problem concerning the aerodynamic stability of the present "stick" configuration, which results in a forward center of pressure (CP) and an aft center of gravity (CG). Such a configuration, aerospace engineers will tell you, is not the best design because the position of the CP and CG will result in increased mechanical loads being imposed on the airframe, although NASA counters these concerns by saying that wind tunnel studies will resolve any stability issues.

Despite the TO problems and the comments of critics, in March 2008 the NASA Agency Program Management Council (APMC) gave unanimous approval for the Ares Project to proceed to PDR for the Ares I LV. The APMC approval represented a major programmatic milestone since, in order to receive the approval, the Ares I program had to meet the rigorous requirements and recommendations of the MSFC Center Management Council (CMC) and the Exploration Systems Mission Directorate (ESMD) Program Management Council.

PROPULSION FOR ARES I AND ARES V

Choosing a rocket engine

Propulsion systems usually represent the most technologically challenging and risky aspect of any LV development due to the extremely high operating temperatures, pressures, and mixture ratios involved. To reduce these risks NASA opted for the J-2X engine, whose choice was the result of recommendations made by the original ESAS study.

It was the ELPO that assigned the J-2X Upper Stage Engine Element (USEE) Team to designate a propulsion system for the Ares I US and the Ares V EDS. During ESAS the alternatives considered for the Ares I US included a single expendable variant of the SSME, twin J-2S engines, or a cluster of four new expander engines. Initially, for Ares V, it was decided to redesign an expendable version of the SSME for the Core Stage and to use two J-2S engines for the EDS, but following subsequent engineering studies it was determined that two such engines imposed a weight penalty on the EDS without any significant gain in performance, so the second engine was dropped from the design. Follow-on engineering studies ultimately decided on the RS-68 over the SSME as the main engine for Ares V due to the need to significantly reduce development costs.

J-2 history

The J-2 engine was originally developed by Rocketdyne in the 1960s for the Saturn IB and Saturn V. A simplified version of the engine was also designed by Rocketdyne and designated the J-2S, which, due to design problems, was never flown and not recommended for further development. However, variants of the J-2 design were considered as options for further study and it was an advanced derivative of one of these J-2 variants that resulted in the final J-2X design (Figure 3.8). The choice of the J-2X by the ESAS panel was influenced by several factors but perhaps the most important one was the issue of commonality, one of the stated goals of the Constellation Program.

In May 2006, a commonality assessment examined the commonalities between Ares I and Ares V and concluded that the J-2X interfaces to the stage and main propulsion system should be the same for both vehicles. This decision, when other commonalities such as installation processes and shared sensor and avionics components were considered, led many to believe that one certification program could cover the J-2X for both LVs. A single certification may not be possible, however, due to the different performance requirements and operating environments of the two vehicles, a situation which will probably lead to different engine specifications. The

Figure 3.8. A J-2X engine mounted on the A-1 Test Stand at Stennis Space Center. Between November 2007 and February 2008 tests were conducted on the J-2X's powerpack assembly. The powerpack assembly consists of a gas generator and engine turbopumps originally developed for the Apollo Program. Image courtesy: NASA.

specifications will result in the requirement for only one J-2X start at altitude for an Ares I lunar mission scenario compared with the Ares V requirement of a J-2X ignition at altitude, a burn for eight minutes to place the EDS and LSAM in a stable orbit, and, following a rendezvous with the Orion, a re-start for another five minutes to escape LEO and transfer to lunar orbit. Due to the timeline of NASA's Constellation mission manifest much of Pratt & Whitney Rocketdyne's engineering effort is currently directed at driving development of the J-2X specifications for Ares I. Development of the Ares V J-2X continues but is focused more on resolving some of the problems posed by the CARD requirement of a loiter capability between starting and re-starting the J-2X after a period of up to 95 days. Such a long loiter time poses problems for engineers in the ELPO, not only because they must ensure cryo-coolers for on-orbit propellant management, micrometeorite protection, and additional flight instrumentation, but also because the J-2X engine requires additional igniters to initiate the second start.

At approximately $25 million per engine, the J-2X will cost less than its $55 million SSME counterpart. However, despite the lower cost, the J-2X has a performance advantage over the SSMS since, unlike the SSME, which was designed to start on the ground, the J-2X will feature an air start and vacuum start capability, enabling Ares I to fly the direct insertion profile demanded by the mission architecture.

J-2X development

Thanks to its heritage, the development of the J-2X progressed smoothly both programmatically and technically, achieving several major milestones between the 2006 and 2007 PDRs. In 2006, to accelerate the development of the J-2X, NASA created the Upper Stage Element Office (USEO), which drew upon the experience of several contractor partners and consultants, some of whom had had direct experience with the Apollo program and the original J-2 engine. The Apollo era J-2 engineers, appropriately known as the "Grey-Beard Team", met regularly to discuss previous problems associated with the J-2 and suggested testing procedures, engineering fixes, and developmental lessons that could be applied to the J-2X. Lessons learned from the SSME and EELV programs were also incorporated into the design improvements for the J-2X. These actions led to a procurement strategy and a bottom-up review for the J-2X, which resulted in definition of the engine's design and development approach, identification of project resource requirements, and publication in June 2006 of the planning and design documents necessary to establish a budget and complete the Statement of Work. This milestone was followed in quick succession by documents supporting the development of test facilities to develop, qualify, and certify the J-2X, and a PDR that confirmed that engine requirements were sufficiently established to commence development of subsystems and conceptual engine design.

The next stage in the J-2X development was the transfer of all existing J-2 hardware from the Apollo and X-33 programs to the J-2X program, a process that took place in the summer and autumn of 2006 with the removal of the X-33's turbomachinery and power pack for tear-down, inspection, and subsequent reassignment to the J-2X program at MSFC and Stennis Space Center (SSC). While the J-2 hardware was being reassigned, MSFC engineers completed testing of the J-2X augmented spark igniter (ASI), an assembly of spark plugs, injectors, and an ignition torch that will be used to ignite the LH_2 and LOX propellants in the combustion chamber. Another milestone took place on November 9th, 2006 when the SSC test stand, once used for testing Saturn V stages and SSMEs, was turned over to the Constellation Program for the purposes of testing the J-2X. The occasion was followed shortly after by another landmark event in the same month when NASA completed its first SRR for a manned spacecraft since the development of the Space Shuttle in October 1972. The SRR evaluated the systems of the Orion vehicle, the Ares I and Ares V LVs, and the J-2X which, in its latest configuration, utilizes several heritage components.

Although several elements of the J-2X design have been finalized, the development and testing of the J-2X continues as this book is being written. For example, in December 2007 the first test of the power pack assembly (PPA) for the US engine was achieved, followed by a cold flow "chill" test, which will lead to full-duration hot-fire tests that will generate important information concerning J-2X engine components. Yet to be resolved are issues such as design details of the gas generator, trade studies directed at optimizing the fuel and oxidizer turbopumps, and certain material selections. As these issues are resolved, the J-2X will undergo a series of engine component tests, the first of which was conducted on the powerpacks in the fall of 2007 at SSC. If all goes well, the J-2X will be subject to a CDR in August 2008, an engine systems test in May 2010, main propulsion test article testing in March 2011, followed finally by a Design Certification Review (DCR) in November 2011. After the DCR, Development Engine #1006 will debut the J-2X capabilities in 2013 with the unmanned flight of Orion 3.

J-2X concept of operations

The J-2X fitted to Ares I will ignite approximately 126 seconds after lift-off, following separation of the vehicle's FS at an altitude of 60 km. The engine will burn for 465 seconds, during which time it will expend 137,363 kilograms of propellant and lift Ares I to an altitude of 128 km. Following engine cut-off, the Orion vehicle will separate from the US, upon which Orion's engine will ignite to provide the propulsive power to insert the capsule into LEO where it will rendezvous with either the ISS or the EDS of Ares V, as described previously. The Ares I US with the J-2X engine attached will then de-orbit to splash down in the Indian Ocean. Neither flight element is designed to be reused.

J-2X engineering

The turbomachinery of the J-2X, which consists of the primary power-generating elements such as the gas generator, turbopumps, valves, and feed lines, will utilize hardware initially developed for its sister engine, the J-2S. The operation of this type of rocket engine requires the turbopumps to supply LOX and LH_2 to the main combustion chamber, where the fuel and oxidizer mix and burn at extremely high pressures and temperatures to produce gas, which is forced through the nozzle to produce thrust. The components of the J-2S engine have already been subject to extensive and rigorous testing as part of NASA's 1990s aerospike engine development program which was tasked with developing the engine for the X-33 single-stage-to-orbit (SSTO) Reusable Launch Vehicle (RLV). Sadly, the X-33 project was terminated in March 2001 but, fortunately, in 2006 the engine technology was transferred to support the J-2X development program.

One of the keys to the J-2X's performance is the main injector component, which injects and mixes LOX and LH$_2$ in the combustion chamber, where they are ignited and burned to produce thrust. To date, several hot-fire tests using subscale injector hardware have been conducted at MSFC to investigate design options intended to maximize the performance of the J-2X.

A recent test of the injector occurred in November 2007 when a subscale injector was subjected to 15 hot-fire tests that totaled more than three minutes of main-stage time on the injector. Another significant test occurred in January 2008, when a gas generator ignition (GGI) test was completed using the PPA on the SSC A-1 Test Stand, demonstrating satisfactory ignition and no anomalies. Additional tests in following months culminated in a sixth test on April 3rd, 2008, which was planned for 550 seconds but, due to a LOX valve drifting open, was cut short at 293 seconds. A seventh hot-fire test, conducted 11 days later, was more successful, meeting all test objectives. While the PPA team were conducting hot-fire tests, other J-2X teams were busy conducting PDRs for other engine components such as the helium spin start valve (HSSV) and ancillary valves (AVs), which control valve positions and purges inside the pneumatic control assembly (PCA) of the HSSV. Other J-2X teams are working to meet PDRs for other engine systems as this book is being written with the goal of ensuring the first integrated J-2X engine systems test goes ahead as scheduled in 2010. In the meantime, testing continues at SSC's A-1 Test Stand with the aim of delivering the safest and most cost-effective engine for use on Ares I and Ares V.

ARES V

Ares V overview

Ares V is described in Table 3.5, which also lists some of the more pertinent design requirements mandated by the CARD. Manufactured by the Michoud Assembly Facility (MAF) in Louisiana, the Ares V Core Stage, which measures 10 m in diameter and 64 m in length, is the largest rocket stage ever constructed and is almost as long as the combined length of the Saturn V first and second stages, also manufactured by the MAF.

Ares V Core Stage propulsion

Powering the Ares V Core Stage will be a cluster of five Pratt & Whitney Rocketdyne RS-68 rocket engines, each capable of supplying 700,000 pounds of thrust. The RS-68

Table 3.5. CARD Cargo Launch Vehicle description and requirements [4].

CaLV description   The CaLV provides the heavy-lift capability for the Constellation Program. The CaLV consists of a five-engine Core Stage, two five-segment SRBs, and the Earth Departure Stage (EDS), powered by a J-2X engine (same engine as the CLV upper stage). The EDS serves as the CaLV third stage with a role in injecting the LSAM/EDS stack into LEO staging orbit where the LSAM/EDS and CEV rendezvous and dock. The EDS performs the trans-lunar injection (TLI) burn for the LSAM and CEV after which it is jettisoned.

CaLV requirements

[CA0487-PO]   The CaLV shall limit its contribution to the risk of loss of mission (LOM) for lunar missions to no greater than 1 in 125.
Rationale: The 1 in 125 means a 0.008 (or 0.8%) probability of LOM due to the CaLV during any lunar mission. This requirement is driven by CxP 70003-ANX01, Constellation Need, Goals, Objectives, Safety Goal CxP-G02: Provide a substantial increase in safety, crew survival, and reliability of the overall system over legacy systems.

[CA0049-PO]   The CaLV shall launch LSAM from launch site to the Earth Rendezvous Orbit (ERO) for Lunar Sortie Crew and Lunar Outpost Crew missions.
Rationale: Establishes the CaLV as the launch vehicle to transport the LSAM to the ERO with sufficient remaining propellant to execute the translunar injection burn. The TLI maneuver takes place after CEV docks with LSAM in the ERO.

[CA0391-PO]   The CaLV shall utilize twin Shuttle-derived five-segment SRBs along with a core stage that employs five modified RS-68 engines for first-stage propulsion.
Rationale: The CaLV will take advantage of the flight-proven propulsion systems components developed for the Space Shuttle and Evolved Expendable Launch Vehicles (EELVs). These launch vehicle components, which have supported over 100 Space Shuttle and numerous Apollo missions, have extensive test/flight experience databases available for CaLV designers to leverage. In addition, CaLV designers will be able to leverage the ground processing/production facilities, workforce, and tooling already in place to support Space Shuttle operations.

engine, the most powerful LOX/LH_2 engine in existence, will be modified by a series of upgrades to meet NASA's standards, a process conducted in collaboration with the U.S. Air Force. Once the RS-68 engines are upgraded (Table 3.6), NASA plans to test-fly them on the Delta IV, a strategy designed to reduce technical risk. Given that the reliability of the RS-68 engine is less than that of the SSME, the reduction of risk issue will be an important one to resolve, especially since it is possible that Ares V may be human-rated in the future. The low reliability of the RS-68 was determined by studies evaluating LOM comparisons with the SSME, using a software tool called the Flight-oriented Integrated Reliability and Safety Tool (FIRST), which provided a quantitative risk assessment of the RS-68 compared with the SSME. Unsurprisingly,

Table 3.6. Potential RS-68 upgrades.

Upgrade	Performance improvement
Increase main injector element density	Permits improved propellant mixing and increased combustion efficiency
Material change in engine bearings	Reduces the incidence of corrosion and cracking of bearings
Manifold turbine exhaust ducts that extend from the engine	Mitigates impact upon the five-engine configuration
Thickening the ablative nozzle wall	Increases burn time
Use regenerative cooled nozzle	Increases specific impulse by warming fuel

given the SSME's more than one million seconds of operation from which to extrapolate data, compared with the 39,300 seconds of RS-68 test time, it was inevitable that the reliability of the RS-68 would fall short when compared against the SSME. However, NASA anticipates that once the RS-68 upgrades (Table 3.6) have been implemented and the engine has been subject to testing and certification, it will be adequately reliable for Ares V.

Ares V concept of operations

The current lunar mission architecture will commence with the launch of Ares V from the Kennedy Space Center (KSC) with an Altair lander and EDS under its payload shroud. After Ares V has completed one orbit, Ares I, carrying the Orion crew exploration vehicle and its four astronauts, will be launched and, following four days on orbital loitering in LEO, the two vehicles will rendezvous and dock and the EDS will be restarted to propel the Orion/Altair combination to the Moon.

Ares V development problems

At the time of writing, Ares V still doesn't have the lift required to launch astronauts and their hardware to the Moon and engineers are studying several options for adding the 11-tonne shortfall, needed to meet the designed 75.1-tonne lift capability including margin. Another problem faced by engineers is the uncertainty over the mass of the flight components such as the Altair Lander and the Orion Crew Vehicle, each of which are gaining weight as upgrades are installed. Some of the options suggested by engineers include adding a sixth RS-68 engine to the Ares-V LOX/LH$_2$ Core Stage, extending the length of the solid-fuel boosters, or even combining both solutions. Other options include scrapping the reusable steel booster casings used in the Shuttle program and using lighter composite casings capable of containing higher operating pressures. Such a solution would boost the estimated performance by 6.1

tonnes to 64 tonnes which the Ares V can produce in its present configuration. A sixth engine and the addition of an inert spacer to stretch the solid-fuel boosters adds another 5 tonnes in lift performance which would bring total lift capability to the design requirement. Compounding the problem of how many modifications engineers can make to the Ares V is the restriction that the vehicle must fit inside the Vehicle Assembly Building (VAB), whose ceiling height represents the height limit of Ares V.

ARES V ELEMENTS

Earth Departure Stage

The Ares V US, more commonly referred to as the EDS, will be powered by the same J-2X engine that will power the Ares I US. For Ares V missions the first time the J-2X will be ignited will be 325 seconds after launch at an altitude of approximately 120 km, following separation of the Ares V FS from the EDS and LSAM. Following a burn of 442 seconds, during which the J-2X will use 131,818 kilograms of propellant, the EDS and LSAM will be placed in LEO. The second ignition of the Ares V J-2X will occur once the Orion Crew Vehicle, delivered to LEO by the Ares I, has docked with the EDS and LSAM (Figure 3.9). Once these flight elements are mated the J-2X will fire for a second time for a period of 442 seconds, which will provide sufficient power to accelerate the mated vehicles to the escape velocity required for the Orion–EDS–Altair combination to break free of Earth's gravity and enter a trajectory known as translunar injection (TLI). Once the mated vehicles arrive in lunar orbit, the Orion–Lunar Lander combination will jettison the EDS and its J-2x engine and perform a maneuver that will send these two flight elements into orbit around the Sun.

ALTAIR/LUNAR SURFACE ACCESS MODULE

Altair design history

The process of defining Altair's configuration required consideration of a number of lander concepts generated by the Constellation Program's Advanced Projects Office Lunar Lander Preparatory Study (LLPS), initiated in early 2006. The first phase of the study generated 37 lander configurations, some of which are described here.

MSFC suggested the lander concept, a vertical lander with a sidemount ascent stage, designed to be used as an airlock, and a cylindrical surface habitat positioned in the center of the descent tanks. In contrast to MSFC's design, the Jet Propulsion Laboratory (JPL) submitted a proposal for a split habitat crew lander with a minimally sized descent/ascent habitat comprising a vertical lander featuring two hydrogen tanks which would be converted to habitats following landing. The Glenn Research Center (GRC) submission was a simplified split descent concept that

Figure 3.9. Earth Departure Stage and Lunar Surface Access Module docked in low Earth orbit. Image courtesy: Boeing Company.

featured a drop stage lander and a single reusable cryogenic stage. In contrast to the other designs, the GRC concept was designed to leave behind only some landing gear and ancillary systems since the habitat would serve as both the lander and ascent vehicle.

A combined GSFC–JSC–GRC concept proposed some innovative ideas for transferring crewmembers and cargo from the descent stage deck to the surface thanks to a novel vehicle configuration design. The main element in the GSFC–JSC–GRC lander was the Minimum volume Ascent Vehicle (MAV) which was intended to not only serve as sleeping quarters and extended living space while on the lunar surface, but also to transport crew to and from the surface.

From the broad spectrum of lander configurations developed by the NASA field centers, six spacecraft designs were chosen for further evaluation by a team of engineers at the NASA Langley Research Center (LRC), which was tasked with developing two different lander concepts for the second phase of the LLPS. The two vehicles chosen were the Descent Assisted, Split Habitat (DASH) Lander Vehicle and the Cargo Star Horizontal Lander (CSHL), each with the capability of ferrying four crewmembers to and from the lunar surface in a minimum-volume pressurized module.

DASH Lander Vehicle concept

The DASH Lander Vehicle was composed of three modules, the first being the Lander Module (LM), which contained the crew and all critical lander subsystems such as propulsion and avionics. The propulsion system utilized two engines derived from the SSME. The second module was the Payload Module (PM), which provided a platform for payloads, such as pressurized habitats and outpost infrastructure to be transported to the lunar surface. The third module was the Retro Module (RM), which served as an in-space braking stage designed to perform the LOI maneuver, in addition to 90% of the delta-V required for descent to the lunar surface.

Structurally, the core of the DASH Lander Vehicle comprised two octagonal truss platforms connected to node fittings to which all major vehicle components and subsystems were attached. A similar structural configuration was employed in the construction of the Lander Module's pressurized Transport Habitat, which was attached to the central nodes of the upper truss platform and stabilized using truss members, a design intended to reduce the loads upon the TH during TLI and LOI burns. At the base of the TH was a short tunnel that passed through the upper truss platform to permit access to the SH from the TH. Access to the lunar surface was either via an inflatable airlock attached to one side of the SH or in an emergency via the LIDS hatch on the TH.

Cargo Star Horizontal Lander concept

The CSHL was designed to carry cargo and crew to a lunar outpost site. Sized for a crew of four and a seven-day stay, the CSHL was designed around a large central cargo bay positioned on the Descent Stage (DS), while a pressurized SH, used for lunar surface operations, was stowed in the cargo bay. Connected to the SH by a tunnel was the Ascent Stage (AS) comprising a horizontal cylinder and central LIDS mechanism for docking with Orion. The tunnel permitted access between the AS and SH following landing.

Current Altair iteration

Although the vehicle that will land astronauts on the lunar surface is still in the design process, the current iteration of the vehicle (Figure 3.10), designated the 711-A, is probably very close to the final design.

Although many people familiar with the Apollo Program point out that Altair shares common features with the Apollo Lunar Module (LM), such as separate descent and ascent stages, Altair's dimensions and technology will make it a very different vehicle (Table 3.7).

Operational characteristics

The design requirements of the LSAM are described in Table 3.8, which also lists some of the CARD mission-critical roles the LSAM must fulfill.

Figure 3.10. Artist rendering of Altair on the surface of the Moon. Image courtesy: NASA.

Table 3.7. Altair compared with Apollo Lunar Module.

Feature	Apollo Lunar Module	Altair
Crew size	2	4
Surface duration	3 days	7 days (sortie missions) $\geq$210 days (outpost missions)
Landing site capability	Near-side, equatorial	Global
Stages	2	2
Width at tanks	4.22 m	7.5 m
Width at footpads	9.45 m	14.8 m
Crew module pressurized volume	6.65 m^3	31.8 m^3
Ascent stage mass	4,805 kg	10,809 kg
Ascent stage engines	One UDMH–NTO	One LOX–CH$_4$
Ascent engine thrust	15.6 kN	44.5 kN
Descent stage mass	11,666 kg	35,055 kg
Descent engine thrust	44.1 kN	4×66.7 kN

Table 3.8. CARD Lunar Surface Access Module description and requirements [4].

LSAM description The LSAM transports cargo to LLO and crew and cargo from LLO to the lunar surface and back. The LSAM is intended to support lunar DRMs. LSAM may be configured with and without crew. The uncrewed configuration transports significant cargo in support of extended lunar outpost missions and does not include an ascent capability from the lunar surface. The uncrewed/cargo version of the LSAM, without ascent capability, may be used to store supplies or waste upon completion of its cargo delivery mission. The LSAM is capable of using its Descent Stage to insert itself and CEV into LLO and carry crew or cargo to the lunar surface. For crewed lunar sortie configurations, the LSAM serves as the crew's home for up to 7 days and uses an Ascent Stage to return them to LLO. The Descent Stage serves as the launch platform for the Ascent Stage and is discarded on the lunar surface. The Ascent Stage is jettisoned prior to CEV trans-Earth injection (TEI) from LLO.

LSAM requirements	
[CA0504-PO]	The LSAM shall limit its contribution to the risk of loss of mission (LOM) for a lunar sortie crew mission to no greater than 1 in 75. *Rationale:* The 1 in 75 means a 0.013 (or 1.3%) probability of LOM due to the LSAM during any lunar sortie crew mission. This requirement is driven by CxP 70003-ANX01, Constellation Program Plan, Annex 1: Need, Goals, and Objectives (NGO) Safety Goal CxP-G02: Provide a substantial increase in safety, crew survival, and reliability of the overall system over legacy systems.
[CA0062-PO]	The LSAM shall return at least 100 kg of payload from the lunar surface to lunar rendezvous orbit (LRO) during each crewed lunar mission. *Rationale:* The LSAM returned mass must include the 100 kg of payload specified by ESMD in addition to the crew and flight crew equipment. This requirement applies to each crewed lunar mission and is needed to size the LSAM Ascent Stage.
[CA5236-PO]	The LSAM shall perform aborts from post-TLI until lunar landing for lunar sortie and lunar outpost crew missions. *Rationale:* The Constellation architecture will have abort capabilities for all mission phases. The LSAM will support aborts during several phases that begin following LSAM–EDS separation, when the LSAM becomes an active vehicle, and end when the LSAM lands on the Moon. During translunar coast, the LSAM may provide abort capabilities while docked to the CEV. Following undocking in lunar orbit, the LSAM will support aborts during lunar powered descent. Abort opportunities end at lunar landing, transitioning to opportunities for early return.
[CA3200-PO]	The LSAM shall utilize a liquid hydrogen/liquid oxygen (LH_2/LOX) Descent Stage propulsion system that can be throttled. *Rationale:* The operational concept described in CxP 70007, Constellation Design Reference Missions and Operational Concepts Document, leads to the LSAM performing multiple functions including lunar orbit insertion (LOI), lunar descent, and lunar landing. To execute these functions, the LSAM Descent Stage propulsion system requires the flexibility to throttle the engine to control the propellant usage and engine performance. A LH_2/LOX propulsion system can be throttled and provides the delta-V efficiency needed with a system mass that is within the Launch Vehicle capabilities.

In its final iteration, Altair will be capable of transporting four crewmembers to and from the lunar surface, provide crews with global access capability, allow astronauts to return to Earth at any time during a mission, and be capable of landing 20 tonnes of mission-dedicated cargo. In order to provide these capabilities, Altair will have variants for sortie, cargo, and crewed outpost missions and will feature characteristics similar to those in Table 3.9.

Altair concept of operations

Mounted atop an EDS, Altair will be launched to LEO aboard Ares V. Once mated, Altair and the EDS must then survive a loiter period of up to 95 days in LEO and wait for the four crewmembers in Orion, which will be launched atop Ares I. Orion will then rendezvous with Altair and the vehicles will mate using an androgynous Low-Impact Docking System (LIDS), originally developed for the X-38 Program. The EDS will then perform a translunar injection (TLI) burn to place the Altair–Orion configuration on an Earth–Moon trajectory. Upon completion of the TLI burn, the EDS is discarded. When the Altair–Orion configuration arrives at the Moon, Altair will perform a lunar orbit insertion burn, and following a loiter period in low lunar orbit (LLO), the crew will transfer aboard Altair and descend to the lunar surface (Figure 3.11).

Following either a seven-day sortie mission or a longer duration outpost mission the crew boards Altair's Ascent Stage and launches to LLO for rendezvous with Orion. When Altair launches from the lunar surface it must perform a number of maneuvers before rendezvousing with Orion, waiting in a 100 km by 100 km parking orbit. The first phase of Altair's ascent profile begins with a 100-meter vertical rise [2], lasting approximately ten seconds. The vertical rise phase is followed by a single-axis rotation (SAR) maneuver, which alters Altair's rotation, attitude, angular velocity, and angular acceleration in preparation for the next phase, called the powered explicit guidance (PEG) phase that delivers Altair to the desired orbit [6–8]. Following main engine cut-off (MECO), Altair coasts for approximately ten minutes before commencing a series of burns and maneuvers that ultimately lead to rendezvousing with Orion. To calculate Altair's burns and maneuvers, engineers use depth of field (DoF)[1] simulation and modeling software called simulation and optimization of rocket trajectories (SORT), originally developed for defining and optimizing the ascent of the Space Shuttle. Another tool used by mission planners is the Flight Analysis System (FAS), consisting of several interlinked modules that can consider factors such as lighting and ground tracks and how they affect a spacecraft.

Using these tools mission planners have been able to refine Altair's rendezvous maneuvers, occurring after the initial ten-minute coast following orbital insertion. As it may not be possible for Altair to rendezvous with Orion after just one orbit, a one-orbit and two-orbit rendezvous sequence has been calculated.

---

[1] Depth of field is the amount of distance between the nearest and farthest objects that appear in acceptably sharp focus as seen by a camera lens. Depth of field is used in computer graphics as a creative element.

Table 3.9. Altair variant characteristics.

Lander performance	
Crew size	4
LEO loiter duration	14 days
Launch shroud diameter	8.4 m
Lander design diameter	7.5 m
Surface stay time	7 days (sortie), 180 days (outpost)
Launch loads	5g axial, 2g lateral
Crewed Lander mass (launch)	45,586 kg
Crewed Lander mass @ TLI	45,586 kg
Crew Lander payload to surface	500 kg
Crew Lander deck height	6.97 m
Cargo Lander mass (launch)	53,600 kg
Cargo Lander mass @ TLI	N/A
Cargo Lander payload to surface	14,631 kg
Cargo Lander height	6.97 m
Crew Lander LOI delta-V capability	891 m/s
Cargo Lander LOI delta-V capability	889 m/s
Crew descent propulsion delta-V capability	2,030 m/s
Cargo descent propulsion delta-V capability	2,030 m/s
TCM delta-V capability (RCS)	2 m/s
Descent orbit insertion capability (RCS)	19.4 m/s
Descent and landing reaction control capability	11 m/s
Ascent delta-V capability	1,881 m/s
Ascent RCS delta-V capability	30 m/s

Vehicle concept characteristics	
Ascent Module	
Diameter	2.35 m
Mass @ TLI	6,128 kg
Main engine propellants	N_2O_4/MMH
# Main engines/type	1/derived OME/RS18 pressure-fed
Usable propellant	3,007 kg
Main engine I_{SP} (100%)	320 s
Main engine thrust (100%)	5,500 lbf
RCS propellants	N_2O_4/MMH
# RCS engines/type	16/100 lbf
RCS engine I_{SP} (100%)	300 s
Airlock	
Pressurized volume	7.5 m^3
Diameter	1.75 m
Height	3.58 m
Crew size	2+
Descent Module (crewed)	
Mass @ TLI	38,002 kg
Main engine propellant	LOX/LH$_2$
Usable propellant	25,035 kg
# Main engines I_{SP} (100%)	1/RL-10 derived
Main engine I_{SP} (100%)	448 s
Main engine thrust (100%)	18,650 lbf
RCS propellants	N_2O_4/MMH
# RCS engines/type	16/100 lbf each
RCS engine I_{SP} (100%)	300 s

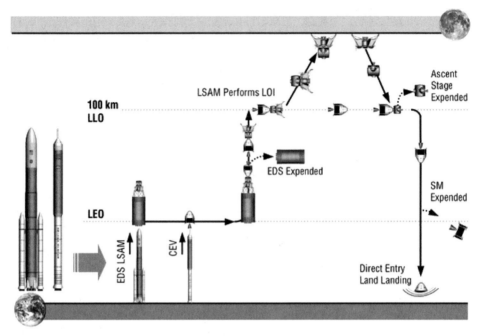

Figure 3.11. Altair concept of operations. Image courtesy: NASA.

One-orbit rendezvous maneuver

Upon completion of the coast phase, Altair will perform a corrective combination maneuver designed to create a differential altitude between it and Orion at the first closest point of approach. When Altair is at its farthest point from the Moon during its orbit the terminal phase initiation (TPI) burn will occur, whose timing and velocity changes are carefully planned to place Altair in the vicinity of Orion, following a designated orbital transfer angle. To optimize the transfer angle, the crew will be able to perform additional mid-course correction burns during the transfer, which will end with a series of small braking maneuvers to slow Altair's approach rate to Orion. The events and timings of the one-orbit rendezvous sequence are shown in Table 3.10.

Two-orbit rendezvous maneuver

For the two-orbit maneuver the first two phases remain the same. Altair coasts to a point in its orbit that is farthest away from the Moon's center[2] and then performs a corrective combination maneuver to create a differential altitude between Altair and Orion at the first closest point of approach. The next maneuver is termed the coelliptic maneuver, which places Altair on an intercept trajectory with Orion at the required altitude. Following the coelliptic maneuver, Altair coasts for 30 minutes

---

[2] This point is termed *apolune*, whereas the point on an orbit that is nearest the Moon is termed *perilune*.

Table 3.10. Altair–Orion one-orbit rendezvous sequence [9].

Trajectory event	Phase elapsed time (dd:hh:mm:ss)	Delta velocity magnitude of translational maneuver (m/s)	Farthest excursion/ closest point of approach (km)
Lift-off	00:00:00:00		74/15
Insertion	00:00:07:01		75/15
Corrective combination	00:00:17:01	1	101/74
Terminal phase initiation	00:01:03:01	21	100/100
Terminal phase finalization	00:01:45:34	8	100/100
Docking	00:03:03:00		100/100

before initiating the TPI burn, a coast period that places the vehicle in the vicinity of Orion. In common with the one-orbit rendezvous maneuver, the two-orbit variant permits the crew to perform mid-course correction burns during the transfer time. The two-orbit rendezvous maneuver is summarized in Table 3.11.

Launch window

Altair's launch windows will occur approximately every two hours, assuming a 26°

Table 3.11. Altair–Orion two-orbit rendezvous sequence [9].

Trajectory event	Phase elapsed time (dd:hh:mm:ss)	Delta velocity magnitude of translational maneuver (m/s)	Farthest excursion/ closest point of approach (km)
Lift-off	00:00:00:00		
Insertion	00:00:07:01		75/15
Corrective combination	00:01:03:01	14	75/74
Coelliptic maneuver	00:02:00:01	1	75/75
Terminal phase initiation	00:02:40:01	8	101/74
Terminal phase finalization	00:03:22:34	8	101/100
Docking	00:04:40:00		101/100

latitude launch site, but due to perturbations to Orion's orbit the required delta-V costs will vary. However, when mission planners calculate launch windows they must not only take into consideration latitude and orbital perturbations but also factors such as the ascent *plane window* and rendezvous *phase window* [9].

The plane window is basically a range of times during which Altair may launch and still be able to "chase down" Orion in lunar orbit, whereas the phase window is a range of angles (defined at the spacecraft's ascent engine cut-off) that a spacecraft may launch in order to facilitate a rendezvous with an orbiting vehicle. As can be seen in Tables 3.10 and 3.11, there is no phasing burn in either the one-orbit or two-orbit rendezvous sequences, which means there is no rendezvous phase window associated with these sequences, which in turn means that Altair must launch on time. If Altair does not launch on time the crew have the option of implementing a phase window of between 30 s and 40 s, but they will only be able to do this at the cost of increasing delta-V by the order of between 10 m/s to 20 m/s. Given the two-hour orbit of Orion and the location of Altair at latitude 26°, an optimum launch window in terms of total delta-V will occur every two hours.

In addition to refining lunar launch windows, mission planners also analyzed variables that affect ascent performance such as thrust-to-weight ratio, orbit perilune, and orbit apolune, in an attempt to optimize and refine the design of Altair's trajectory and flight mechanics. For example, to investigate the effects of perilune, mission planners chose 15.24 km as the lowest safe altitude perilune [3] and then calculated the changes in delta-V based on the variation with the perilune of the orbit at insertion. A similar calculation was then performed for apolune and the results compared. This comparison revealed that the effect upon ascent performance of changing the perilune was a significant 2.4 m/s of delta-V per kilometer, whereas the cost of changing to apolune was only 0.22 m/s per kilometer.

Abort methods

"The Constellation Architecture shall provide the capability to perform an expedited return of the crew from the surface of the Moon to the surface of the Earth in 120 hours or less after the decision to return has been made."

Constellation Architecture Requirements Document requirement CA0352-HQ

The CARD requirement demands that a crew have the means to depart any lunar landing location before the nominal end of a mission, a capability usually termed an *anytime return* capability. This section describes this capability and other abort options available to the crew in the event of an off-nominal event.

Anytime return

During a nominal seven-day sortie mission it may be necessary for Altair to launch from the Moon and rendezvous with Orion at any time, an option that poses problems for mission planners as Orion will not pass directly over Altair's launch site during the surface duration. Another complication is that Altair may be located anywhere on the lunar surface. One method of inserting Altair into the correct orbit for rendezvousing with Orion is to steer the vehicle during ascent, thereby achieving the required inclination of Orion's orbit, although this method assumes that Orion's orbit remains unchanged by the perturbative effect of the lunar gravity field.

Rendezvous following powered descent initiation abort

If a systems failure occurs during Altair's powered descent to the lunar surface, crewmembers will have two options available to them. If a systems failure occurs late in the powered descent, it will be preferable to land Altair and launch from the lunar surface, rather than abort. In the latter event the crew would wait two hours, which would give them sufficient time to analyze the malfunction, perform any system reconfiguration, and allow Orion to complete one orbit of the Moon before performing a nominal ascent and rendezvous. However, if the systems failure occurs early in the powered descent when descent-propulsion-system-to-orbit capability is available, it will be preferable to abort in flight rather than continue the descent, an abort termed a powered descent initiation (PDI) abort, which offers the crew the following three options.

Abort Option 1: Abort at time of powered descent initiation

If the abort is called before the commencement of PDI, the Ascent Module (AM) will separate from the Descent Module (DM) and will remain in the post-descent orbit insertion (DOI) orbit and begin a coast period, during which time the crew will make preparations for a five-burn rendezvous maneuver sequence that will commence when the AM reaches its first apolune following the abort. Once the AM reaches its first apolune following the abort call, a height maneuver will be performed which will result in a required height differential between the AM and Orion, thereby boosting the AM to a safer perilune altitude. Half an orbit later, the AM will perform a second burn that will place the vehicle at an angle to prepare it to intercept Orion. One orbit later, the AM will perform the coelliptic maneuver placing the AM in a trajectory designed to intercept Orion. The fourth burn will be the TPI which will place the AM close to Orion and, following the TPI, a series of small braking maneuvers will be performed to slow the AM's approach rate in preparation for docking with Orion (Table 3.12).

Abort Option 2: Abort at time of powered descent initiation plus 120 seconds

Another scenario that may occur is if an abort is called after the PDI burn. In this case the AM will separate from the DM and the crew will perform an insertion burn of approximately 340 seconds, which will result in the AM returning to the post-DOI orbit of 15 km by 100 km. Following insertion to the post-DOI orbit, the AM will wait until the first apolune, at which point a height maneuver will be performed which will create the required differential altitude between the AM and Orion at the next perilune, which will occur one half-orbit later. At perilune, the AM will perform the phasing burn which will prepare the AM for the coelliptic burn scheduled for one

Table 3.12. Powered descent initiation abort sequence [9].

Trajectory event	Phase elapsed time (dd:hh:mm:ss)	Delta V of burn (m/s)	Resultant perilune/ apolune (km)
PDI	00:00:00:00		101/15
PDI abort	00:00:00:00		101/15
Insertion	00:00:00:00	0	101/15
Height maneuver	00:00:59:08	17	101/90
Phasing maneuver	00:01:59:29	29	235/90
Coelliptic maneuver	00:04:03:07	32	90/84
Terminal phase initiation	00:05:05:47	4	105/83
Terminal phase finalization	00:05:54:57	4	106/98
Docking	00:07:00:00		106/98

orbit later. Two hours and one orbit later the AM will perform the coelliptic maneuver, followed one hour later by the TPI maneuver and the necessary braking maneuvers to bring the AM to a stable orbit in preparation for rendezvousing with Orion (Table 3.13).

Abort Option 3: Abort at time of powered descent initiation plus 258 seconds

Another case study abort that mission planners investigated was a PDI abort that occurred 258 seconds after burn initiation, an event that will require the AM to separate from the DM before performing an insertion burn of approximately 691 seconds, which will result in the AM returning to the post-DOI orbit of 15 km by 100 km. In common with the previous abort scenarios, at the first apolune following insertion the AM will perform a height maneuver to create a differential altitude between the AM and Orion at the next perilune. Half an orbit later, the AM will perform a phasing burn to prepare for the coelliptic maneuver, which will occur one orbit later at perilune, placing the AM in an orbit coelliptic to Orion's. After the TPI burn the AM will be placed in the vicinity of Orion in preparation for rendezvous (Table 3.14).

The abort scenarios described indicate that as more of the PDI is complete prior to an abort being initiated, the more delta-V will be required to place the AM back into lunar orbit. Another feature is the length of time that will be required to complete the rendezvous with Orion, which will probably be between four orbits (seven hours) and five orbits (nine hours).

Table 3.13. Abort 120 seconds following powered descent initiation [9].

Trajectory event	Phase elapsed time (dd:hh:mm:ss)	Delta V of burn (m/s)	Resultant perilune/ apolune (km)
PDI	00:00:00:00		101/15
PDI abort	00:00:02:00		18/1,400
Insertion	00:00:05:40	812	101/15
Height maneuver	00:01:01:44	17	101/90
Phasing maneuver	00:02:02:05	24	213/90
Coelliptic maneuver	00:04:04:40	28	90/84
Terminal phase initiation	00:05:07:14	4	105/83
Terminal phase finalization	00:05:56:24	4	106/98
Docking	00:07:00:00		106/98

ARES V PROGRESS

It will be a number of years before Ares V starts flying lunar missions. However, due to the lengthy process of designing and developing a new heavy-lift LV system the Ares V team is already busy conducting design and analysis work, participating in various reviews and study forums and developing safety, reliability, cost, and technical performance metrics. For example, in the summer of 2006, an RSRB was fitted with 117 instrument channels and subjected to a two-minute static fire test at the ATK Launch Systems Test Facility.

Accelerating operational status

In recent panels, press conferences, and symposia there has been talk of a push for a post-election increase in NASA funds to reduce the gap in America's human space-flight capability by almost two years or to eliminate the gap altogether in a long-shot proposal to keep one Shuttle flying every six months or so. The figure that keeps on cropping up is an additional $1 billion in Fiscal 2009 followed by another $1 billion in Fiscal 2010, sums that would allow NASA to advance the March 2015 first operational flight of the Orion to September 2013. Another argument for the increase in funds is that by front-loading NASA with money, problems such as TO could be resolved sooner and by flying a Shuttle twice a year the problem of job losses

Table 3.14. Abort 258 seconds following powered descent initiation [9].

Trajectory event	Phase elapsed time (dd:hh:mm:ss)	Delta V of burn (m/s)	Resultant perilune/ apolune (km)
PDI	00:00:00:00		101/15
PDI abort	00:00:04:18		5/1,700
Insertion	00:00:11:31	1,848	101/15
Height maneuver	00:01:07:46	17	101/90
Phasing maneuver	00:02:08:06	3	116/90
Coelliptic maneuver	00:04:07:53	7	91/84
Terminal phase initiation	00:05:10:33	4	106/84
Terminal phase finalization	00:05:59:44	4	106/98
Docking	00:07:00:00		106/98

associated with the loss of Shuttle capability would be negated. If the option to fly one Shuttle was approved it has been suggested that just one bay of the Vehicle Assembly Building (VAB), one mobile crawler, and one pad be retained and that the remaining orbiters be used as hangar queens and cannibalized for parts no longer available from the dwindling Shuttle supplier base.

REFERENCES

[1] Advanced Transportation System Studies Technical Area 2 (TA-2). *Heavy Lift Launch Vehicle Development Contract*, NAS8-39208, DR 4, Final Report. Lockheed Martin Missiles & Space for the Launch Systems Concepts Office of the George C. Marshall Space Flight Center (July 1995).
[2] Bennett, F.V. *Lunar Descent and Ascent Trajectories*, MSC Internal Note No. 70-FM-80. Marshall Space Flight Center, Houston, TX (April 21, 1970).
[3] Bennett, F.V.; Price, T.G. *Study of Powered Descent Trajectories for Manned Lunar Landing*, NASA TN D-2426. NASA, Washington, D.C. (August 1964).
[4] NASA. *Constellation Architecture Requirements Document*, NASA CxP 70000. Baseline. NASA, Washington, D.C. (December 21, 2006).
[5] NASA. *Exploration Systems Architecture Study: Final Report*, NASA-TM-2005-214062. NASA, Washington, D.C. (November 2005).
[6] Ives, D.G. *Shuttle Powered Explicit Guidance Miscellaneous Papers*, JSC-28774. Aeroscience & Flight Mechanics Division (November 1999).

[7]   Jaggers, R.F. *Shuttle Powered Explicit Guidance (PEG) Algorithm*, JSC-26122. Rockwell Space Operations Company (November 1992).

[8]   McHenry, R.L.; Brand, T.J.; Long, A.D.; Cockrell, B.F.; Thibodeau, J.R. Space Shuttle ascent guidance, navigation, and control. *Journal of the Astronautical Sciences*, **XXVII**(1), January/March 1979.

[9]   Sostaric, R.R.; Merriam, R.S. Lunar ascent and rendezvous trajectory design. *31st Annual AAS Guidance and Control Conference, Breckenridge, Colorado, February 1-6, 2008*, AAS 08. American Astronautical Society, Breckenridge, CO.

4

Designing the Crew Exploration Vehicle

Men go into space to see whether it is the kind of place where other men, and their families and their children, can eventually follow them. A disturbingly high proportion of the intelligent young are discontented because they find the life before them intolerably confining. The Moon offers a new frontier. It is as simple and splendid as that.

Editorial on the Moon landing, *The Economist*, 1969

In September 2010, the Space Shuttle will be retired and its role of transporting astronauts to and from the International Space Station (ISS), the Moon, and ultimately Mars will be met by the Orion Crew Exploration Vehicle (CEV) due to be launched in 2013.

Orion (Figure 4.1), which is being built by Lockheed Martin, comprises four functional modules. On top of the Ares I Launch Vehicle (LV) sits the Spacecraft Adapter (SA) which serves as the structural transition to the Ares I LV. Above the SA is the unpressurized Service Module (SM) which provides propulsion and electrical power to the pressurized Crew Module (CM)/Orion, designed to transport either crew or cargo. Above Orion sits the Launch Abort System (LAS) which provides the crew with an emergency escape system during launch. Although the modular design appears familiar to those who remember the Apollo program the vehicle configuration takes full advantage of all the technology the 21st century has to offer.

This chapter describes the design and development of each functional module. However, given the complexity of designing and developing a new manned spacecraft, the focus is necessarily directed at those issues as they apply to Orion. To that end, before describing Orion's design and development it is useful to understand the contractor studies that led to the selection of the company chosen to design Orion.

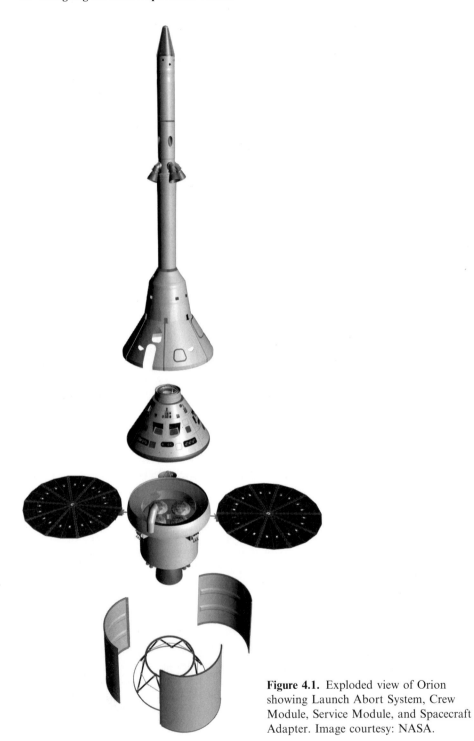

Figure 4.1. Exploded view of Orion showing Launch Abort System, Crew Module, Service Module, and Spacecraft Adapter. Image courtesy: NASA.

CREW EXPLORATION VEHICLE CONTRACTOR STUDIES

On September 4th, 2004, NASA, under the stewardship of Sean O'Keefe, announced their selection of companies for the initial six-month CEV studies (Tables 4.1 and 4.2, see p. 88). NASA's original plan was to whittle down the number of companies to two, who, at the end of the six-month period, would develop competing CEV designs and demonstrate them in unmanned flight tests to be conducted in 2008. Following completion of the unmanned tests NASA would choose a single prime contractor who would develop, test, human-rate, and deploy a CEV capable of fulfilling NASA's requirements.

Andrews Space Inc. Crew Exploration Vehicle

Andrews Space was formed in 1999 and employed a staff of only 30. Led by husband-and-wife team Jason Andrews and Marian Joh, Andrews was a company with revenues in the millions of dollars compared with the billion dollar incomes of the mega-teams such as Boeing and Lockheed.

The Andrews Space Inc. CEV (Figure 4.2, see p. 88) adopted the Apollo CM re-entry shape and combined it with a forward cylindrical mission module and an aft Orbital Transfer Vehicle (OTV) for propulsion. The capsule was 4.5 m in diameter, providing sufficient volume to support a crew of four during lunar missions and a crew of six for trips to and from the ISS. The development cost was estimated to be $6.191 billion, with a unit cost of $224 million.

Table 4.1. Contractors chosen for CEV development studies.

Contractor	Location	Contract base value ($)
Andrews Space Inc.	Seattle, Washington	2,999,988
Draper Labs	Cambridge, Massachusetts	2,988,083
Lockheed Martin Corp.	Denver, Colorado	2,999,742
Northrop Grumman Corp.	El Segundo, California	2,958,753
Orbital Science Corp.	Dulles, Virginia	2,998,952
Schafer	Chelmsford, Massachusetts	2,999,179
The Boeing Co.	Huntingdon Beach, California	2,998,203
t-Space	Menlo Park, California	2,999,732

Figure 4.2. Andrews Space Inc. mock-up of Crew Exploration Vehicle proposal. Image courtesy: Andrews Space Inc.

Draper Labs

Draper's design approach was to work backward from defining the Mars mission. The company then applied these elements to the lunar landing mission, a method that produced some unusual mid-term and final assessments.

Draper's mid-term concept envisaged the CEV as an integral biconic lifting body, 5 m in diameter, capable of providing a crew of four with a habitable volume of 22 cubic meters. By the time the final report was published, Draper's CEV had morphed into a squat ballistic capsule that provided a crew of five with a habitable volume of 12.8 cubic meters. One drawback of Draper's CEV design was that the vehicle was not designed to be fitted with an airlock, requiring the capsule to be depressurized prior to conducting EVAs. However, this disadvantage was compensated for by the CEV's modular subsystems which conferred a high reuse capability.

Lockheed Martin Corporation

Lockheed Martin had previously been awarded $1.2 billion to build the X-33 hypersonic plane, a vehicle destined to replace the Space Shuttle, but the project was canceled when cracks were detected in the X-33's experimental fuel tanks.

Given Lockheed's extensive lifting body experience, it was no surprise that the first CEV design proposed by one of the United States leading spacecraft manu-

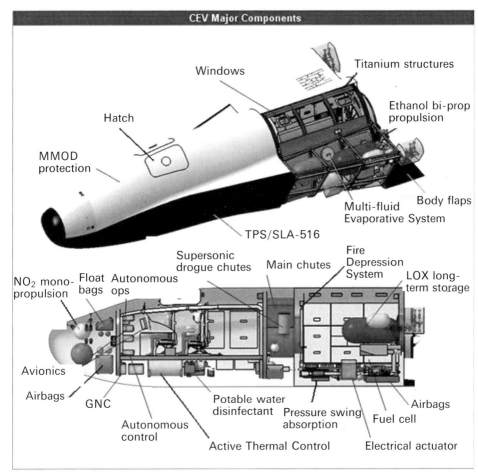

Figure 4.3. Cutaway of Lockheed Martin's lifting body proposal for the Crew Exploration Vehicle. Image courtesy: Lockheed Martin.

facturers was a lifting body. Lockheed argued that the airplane shape of its proposed CEV would make it easier to navigate during Earth re-entry than capsule-shaped vehicles such as Apollo and Soyuz. However, Lockheed's first CEV iteration was rejected by NASA, forcing the company back to the drawing board to design a ballistic capsule that bore more than a passing resemblance to the Soyuz spacecraft used by the Russian Space Agency (Figure 4.3).

Northrop Grumman Corporation

Originally, Northrop Grumman had designed a CEV concept that comprised a three-module spacecraft consisting of a Service Module (SM), Re-entry Module (RM), and a Habitat Module (HM), but this configuration was designed for Northrop's lunar architecture which required transit to the L1 Moon–Earth Lagrangian point. However, the L1 scenario wasn't NASA's preferred architecture, so Northrop reverted to a design similar to the original Apollo configuration.

Orbital Sciences Corporation

Orbital Sciences Corporation (OSC) had gained previous spacecraft design experience when it was a lead player in the X-34 Orbital Space Plane (OSP) project, which was canceled before any hardware was flown.

OSC's Apollo-derived modular CEV concept comprised a CM, a Space Exploration Module (SEM), and a LAS. The CM, which Orbital estimated would cost $133 million, provided habitation for four crewmembers from launch to either Earth or lunar orbit and a return to Earth. The SEM provided power to the CM and the propulsive capability to transfer the CM or lunar lander from low Earth orbit (LEO) to lunar orbit and return to Earth. The LAS comprised a tractor rocket system that provided the crew with an abort capability up to 60 km altitude.

Schafer

Schafer's CEV was envisaged as being capable of supporting four crewmembers for 17 days. The CEV would be completely reusable and carry sufficient propellant to enable a 3.07 km/s delta-V necessary for trans-Earth injection (TEI) directly from the

Table 4.2. Contractors CEV design and mission plan summary.

Contractor	CEV concept	Diameter (m)	Structure mass (kg)	Crew	RV form
Andrews Space Inc.	MM + CM + SM	4.50	2,639	6	Apollo
Draper Labs	Integral	5.00	8,000	6	Soyuz
Lockheed Martin Corp.	MM + CM + SM	4.20	20,000	4	Lifting body
Northrop Grumman Corp.	MM + CM + SM	5.00	21,000	4	Soyuz
Orbital Science Corp.	Integral	5.00		4	Apollo
Schafer	Integral			4	Discoverer
The Boeing Co.	MM + CM + SM	4.50	20,000	4	Apollo
t-Space	Integral	4.20	5,000	4	Discoverer
NASA	Integral			6	Apollo

lunar surface. Other advantages of Schafer's CEV design included an "anytime abort capability" to return to Earth and the capability of returning up to 500 kg of payload in addition to the crew.

The Boeing Company

Boeing's 26-tonne CEV comprised an Apollo-type capsule, a 14.5-tonne SM, and a pressurized Mission Module (MM), which would be capable of transporting four crewmembers. For single-launch Earth orbit missions the MM would be stowed under the CEV during launch. In orbit the MM would dock with the CEV, whereas during lunar landing missions the MM would be launched separately with the Lunar Lander. In this latter mission mode, the MM would serve as a lunar surface cabin.

Transformational Space

The Crew Transfer Vehicle (CXV)/Discoverer concept proposed by Transformational Space (t/Space) and Burt Rutan's Scaled Composites was a scaled-up iteration of the United States Air Force's Corona capsule. Since Corona's aerodynamic and flight characteristics were well known thanks to data obtained from more than four hundred re-entries Rutan's team were able to design an extremely safe and versatile vehicle.

The CXV/Discoverer was designed for the sole purpose of delivering crew to and from LEO. Thanks to its shape, the capsule would automatically right itself during its descent through the atmosphere, no matter what the aerodynamic loads or orientation, a design factor that conferred a significant safety benefit upon the crew. Furthermore, due to the capsule's CG being slightly offset from the centerline, the descent loads would be maintained below $4g$, an important consideration for deconditioned astronauts returning from a six-month tour of duty onboard the ISS, or from a six-month lunar surface mission. Although concerns were raised concerning the direction of g-forces during the launch and re-entry, t/Space solved the problem by designing a suspended semi-rigid hammock capable of rotating 180° within two seconds, thereby ensuring that crewmembers would be oriented correctly regardless of g-force direction.

Scaled Composites achieved public acclaim on October 4, 2004, when its Tier One System made the first civilian suborbital spaceflight. Based on their final report, t/Space was one of only two companies selected by NASA to receive a $3 million demonstration contract. The award resulted in three successful test drops of a subscale booster conducted in May–June 2005 to prove the efficacy of the unique air launch architecture employed in winning the X-Prize. Having successfully demonstrated their technology, t/Space indicated to NASA they would be able to deliver an uncrewed demonstration flight by summer 2008, to be followed by a crewed flight by 2009. Such a schedule would have eliminated the time gap between the retirement of the Shuttle in 2010 and the initial flight of the CEV which, at that time, was tentatively scheduled for a 2014 flight debut. Furthermore, t/Space's cost of delivering four crewmembers to orbit was estimated at just $20 million per launch, a cost that was less than 10% of NASA's projected budget for developing the CEV. Unfortunately for Rutan and his team, NASA decided to go ahead and focus its resources on its own CEV approach.

Contractor studies outcome

Under NASA Administrator Sean O'Keefe's strategy, CEV development would have proceeded in two distinct phases. Phase 1 would have involved the design of the CEV and a demonstration by contractors that they could safely develop the vehicle, consisting of a suborbital or orbital fly-off known as the flight application of spacecraft technologies (FAST), a kind of competition to see which contractor's vehicle performed the best. At the end of Phase I, a single contractor would be down-selected and Phase II would have commenced, which would have involved final design and construction. Unfortunately for those hoping to see a quick return to the Moon, the strategy proposed by O'Keefe and the proposals based on the contractor studies were not to be realized.

On April 13th, 2005, Mike Griffin was appointed as the new NASA administrator. One of Griffin's first acts was to obtain White House backing to reject all of the contractor proposals and focus instead using NASA's best judgment to construct the elements required for the return to the Moon. One of the reasons for Griffin's decision was based on what he considered to be the unacceptable timeline of the development process, which would leave NASA with a five-year gap between the retirement of the Shuttle and the commencement of CEV flights. Therefore, on June 13th, 2005, NASA announced that Lockheed Martin and the team of Northrop Grumman and Boeing had been selected to build the CEV which would be launched into LEO by the CLV now known as Ares. Ares would be a Shuttle-derived two-stage rocket consisting of a single Shuttle reusable solid rocket booster (RSRB) as the First Stage (FS) and a new Upper Stage (US), using liquid oxygen (LOX)/liquid methane (LH_2) propellants and powered by a single Space Shuttle Main Engine (SSME). On July 12th, 2005, NASA authorized two $28 million eight-month contracts, one to Lockheed Martin and the other to the team of Northrop Grumman and Boeing. During the contract period the companies were to develop designs for the CEV and to demonstrate ability to manage cost, schedule, and risk.

NASA's July 2005 announcement confused many space observers. Many couldn't understand why the agency would choose two companies who had submitted proposals that were essentially upgrades of the Apollo design, rather than opt for the robust and versatile design of Rutan's CXV/Discoverer concept. Furthermore, many of the proposals submitted by the contractors had incorporated the lessons learned as a result of hundreds of millions of dollars of research funded not only by the USAF and the aerospace industry but also by NASA itself. Instead, NASA had chosen to ignore the designs of the contractors and had set up a competition between two aerospace giants, one of whom (Lockheed Martin), although having experience in building unmanned rockets, had never built a manned spacecraft!

The CEV contract

Just over a year later, on August 31st, 2006, NASA announced that the contract to design and develop the CEV had been awarded to Lockheed Martin. The first installment of the contract amounted to $3.9 billion, which will cover design, construction, and testing of the CEV until 2013. By 2013 Lockheed Martin should have built two vehicles. A second influx of $3.5 billion will run from 2009 to 2019 and will cover the building of additional vehicles. August 31st was also notable for the announcement that the CEV would now be known as *Orion*.

The decision went against many space analysts' predictions that Lockheed would lose the contract due to their $912 million X-33 failure. However, the failure may have helped Lockheed win the bid since the X-33 program had provided the company with recent research and development experience in the field of spacecraft design. According to an *Aerospace Daily & Defense Report* summary Lockheed had been awarded the contract on the basis of a superior technical approach and lower and more realistic cost estimates. Other industry analysts speculated the decision was influenced by Lockheed's presence in Washington, D.C, by virtue of an "old boy's network" that exists as a result of the company having former Pentagon and NASA employees on the payroll. Other observers speculated that the contract was awarded to break Grumman–Boeing's monopoly on designing and building manned spacecraft. Needless to say, Lockheed was over the Moon. "We're just tickled, honored to be chosen," said John Karas, Lockheed's vice president for space exploration.

DEFINING AND DESIGNING ORION

The role of the Exploration Systems Architecture Study

One of the roles of the Exploration Systems Architecture Study (ESAS), chartered by NASA, was to assess and define Orion's requirements. The study, conducted between May and July 2005, presented NASA with analysis of and recommendations for Orion's design and an overview of the technologies and potential engineering approaches that could be used to construct the vehicle. The ESAS also provided recommendations to NASA regarding vehicle size, layout, system, and subsystem designs.

Crew Exploration Vehicle design overview

Orion's design is guided by the objectives, requirements, and constraints for the overall mission which, in the case of the Constellation Program, is to land four astronauts on the Moon and return them safely. The requirements and constraints that pertain to Orion's design are described in the *Constellation Architecture Requirements Document* (CARD), described in Chapter 3. For example, Table 4.3 illustrates the CARD description of Orion.

The CARD requirements provide mission planners and engineers with reference points to be used when defining the optimum vehicle design to satisfy the overall

Table 4.3. CARD CEV description and requirements [1].

CEV description   The CEV system consists of a Crew Module (CM), a Service Module (SM), a Launch Abort System (LAS), and a Spacecraft Adapter (SA), and transports crew and cargo to orbit and back. The CEV system will be used in all phases of the Constellation Program. Initially, the CEV transports crew and cargo to and from the ISS and an uncrewed configuration transports pressurized cargo to and from ISS. It will subsequently transport crew and cargo to and from a lunar orbit for short and extended duration missions.

CEV requirements

[CA0056-PO] The CEV shall return the crew and cargo from lunar rendezvous orbit (LRO) to the Earth surface.
Rationale: The Constellation Design Reference Missions and Operational Concepts indicate that the CEV is the Constellation System that will return the crew and cargo to the Earth surface. The CEV includes the propulsion system and propellant to perform the TEI from LRO and any subsequent trajectory correction maneuvers. The CEV includes the heat shield needed for re-entering the Earth's atmosphere and the landing systems needed for return to the Earth surface.

[CA0091-PO] The CEV shall deliver the crew from the Earth surface to the lunar destination orbit (LDO) for crewed lunar missions.
Rationale: The CEV launches on the human-rated CLV. Thus, the CEV is the designated Constellation System, per CxP 70007, Constellation Design Reference Missions and Operational Concepts Document, designed to deliver the crew and cargo from the launch site to Earth rendezvous orbit (ERO) and subsequently from ERO to the lunar destination orbit (LDO).

[CA5312-PO] The CEV shall deliver the crew and pressurized cargo from the Earth surface to the ISS.
Rationale: The requirement is consistent with CxP 70007, Constellation Design Reference Missions and Operational Concepts Document, which indicates that the CEV in the Constellation System is designated to deliver the crew and cargo to the ISS. Design considerations for the CEV must include features that are essential for the crew and cargo to safely launch atop the CLV to the ISS orbit and interface with ISS.

[CA3203-PO] The CEV shall return the crew and pressurized cargo from the ISS to the Earth surface.
Rationale: CxP 70007, Constellation Design Reference Missions and Operational Concepts Document, indicates that the CEV is the Constellation System that will return the crew and cargo to the Earth surface. Design considerations for the CEV need to include features that are essential to returning crew and cargo safely to the Earth surface.

mission goal. This process generates design concepts that satisfy mission requirements, a process often achieved through competition between contractors. Once options have been selected for further development, trade studies are conducted to support the decisions before conceptual designs are developed in order to evaluate feasibility and cost. At this stage, models of the vehicle may be built in order to define vehicle subsystems, mass, and size considerations. Finally, after comparing conceptual designs, the design may be iterated or a baseline will be selected for further development. In the following section each of these steps is explained in more detail.

Generating vehicle concepts

Generating vehicle concepts requires those involved in the design process to be familiar not only with the CARD requirements but also with the general requirements and constraints imposed upon the vehicle design, as summarized in Table 4.4.

Next, the design team considers what are known as *design drivers*, which are usually grouped under the four phases of the mission, namely *Earth ascent*, *space transfer*, *descent and ascent*, and *Earth re-entry*. For example, a design driver for the Earth ascent phase requires engineers to consider crew orientation during launch and the type of emergency egress available to the crew, whereas for the Earth–lunar transfer phase the design team must consider factors such as staging options, radiation protection, and provisions for rendezvous and docking. Using the design drivers [3] and CARD's CEV requirements (Table 4.5) as reference guidelines, the team then proceeds with the process of designing the vehicle.

Table 4.4. Requirements and constraints for spacecraft design [2].

Requirements and constraints	Information required
Timeframe	Start and end of program
Launch mode	Capacity of vehicle, launch environment and launch site
Major maneuvers	Delta-V required
Departure staging	Type and duration of orbit
Destination staging	Use of the ISS
Landing site	Location
Crew	Number
Surface duration	Continuous vs. non-continuous
Payload	Size, mass, pressurized vs. unpressurized
Surface operations	Crew activities; surface transport
Mission duration	Transit times, mission activities
Return mode	Use of the ISS; landing site

Table 4.5. CEV requirements.

Support crew of four from Earth departure to Earth return	Mass between 15 t and 18 t depending on outcome of contract studies
Certification by test to maximum extent possible	Integration capability with Constellation Launch Vehicle (LV) to achieve LEO
Integration capability with Earth Departure Stage to achieve lunar orbit	Integration with Lunar Surface Access Module to achieve lunar surface mission objectives
Maximum use of existing technology	Interface between CEV and Launch System
Abort capability through each flight phase	Open system architecture

ORION TRADE STUDIES

Before Orion's final design was chosen, a number of trade studies were conducted to accurately determine factors such as the most appropriate OML shape, the most effective airlock design, and how much radiation protection to incorporate into the design.

Crew Module shape

Although mission planners and designers had a reasonably good idea of how the CM vehicle should be constructed it was still necessary to be absolutely certain the volume would be sufficient to meet not only the ISS and lunar requirements but also those of the Mars DRM. Other factors engineers needed to assess were the acceleration and entry heating loads in all flight conditions, from nominal to off-nominal to extreme, in order to accurately predict crew survivability and also to satisfy NASA's human-rating requirements.

One aspect of the vehicle shape engineers focused on was monostability, since such a quality would ensure a simple abort technique as the vehicle could attain a stable entry angle of attack without the use of the RCS. The Soyuz vehicle is a monostable vehicle, thanks to its CG located very close to its heat shield, so NASA engineers performed studies attempting to shift the CG to a position that would make the CM monostable.

Having a monostable vehicle was only a part of the re-entry problem, however, since the mission architecture calls for a land landing in the CONUS. Although it was agreed that landing in the CONUS would be a relatively easy task, there are obvious benefits to the crew and recovery teams if the landing can be as accurate as possible. To achieve accuracy, engineers needed to ensure that Orion had a certain lift to drag (L/D) ratio, a vehicle performance quality requiring engineers to perform quantitative trade studies.

The capsule blunt-body shape of Orion was finalized only after considering the advantages and disadvantages of other vehicle classes such as slender bodies, lifting bodies, such as the X-38, and winged vehicles. Although much research had been conducted on winged bodies and lifting bodies, such a vehicle shape was rejected due to the extreme heating encountered upon re-entry and the increased mass due to the requirement for wings, fins, and control surfaces. Once winged bodies had been discounted, the engineers were left to trade the pros and cons of slender bodies and capsules, which they did by evaluating performance and operational differences such as entry heating characteristics, aerostability, crew load directions, and terminal deceleration profiles. For the purpose of analysis, engineers chose a representative class vehicle for each type, the capsule vehicle being an Apollo-shaped CEV and the slender body being a biconic from previous NASA studies.

Once the engineers had decided on which representative vehicles to test they turned their attention to deciding which would be the most appropriate trade tests. As mentioned previously, the re-entry load conditions are important not only because they affect the deceleration of the vehicle but also because of how the loads are encountered by the crew. One trade study therefore calculated aerodynamic loads for the two vehicles, not only during re-entry but also during ascent and abort. The obvious advantage of the blunt body was the load direction, since the logical crew orientation is with their backs parallel to the heat shield, as in this position the body is able to tolerate a higher g-loading. In the biconic shape, however, the crew orientation (and hence, load direction) changes between launch and entry, which means that in the ascent phase of the flight the crew is sitting up, which in turn means that less load may be tolerated. In deciding on vehicle shape, the engineers had to meet the stringent requirements and limits set forth in NASA-STD-3000, Volume III [3], a document that itemizes human systems integration for spacecraft. The crew load limits specified by NASA-STD-3000 meant that engineers had to provide a vehicle that would not expose a crew to a load beyond the maximum permitted during any nominal or off-nominal ascent or entry.

Aerodynamic stability

Orion's shape trade study indicated the blunt body demonstrated better qualities of monostability and directional g-tolerance than the slender biconic but engineers wondered how the biconic would hold up in an assessment of aerodynamic stability.

The aerodynamic stability trade study sought to evaluate the design of Orion as it related to CG location, since it is this value that can have a significant impact upon not only the stability characteristics of the vehicle but also upon heat rates and heat loads, which ultimately affect TPS selection and therefore vehicle mass. By using computational fluid dynamics (CFD), modified Newtonian aerodynamic theory, and wind tunnel analysis, the engineers assessed the L/D ratios of each vehicle to determine how the location of the CG was affected by vehicle trim. One problem associated with the biconic was that the vehicle could only achieve monostability if its CG was placed in a position that was realistically unachievable in terms of the configuration of the crew and internal systems. A second problem was that the location of the biconic's trimline did not remain close to the centerline throughout all phases of flight, a quality that gave the blunt body its very high inherent stability rating.

Ballistic entry

Although the blunt-body design had won the first two trade study assessments, engineers still had to investigate the ability of each vehicle to perform a ballistic re-entry without an active GN&C or power system. In a ballistic entry situation, a spacecraft with high passive stability will likely outperform one with low passive stability since this quality permits a spacecraft to reorient itself to a nominal attitude from an off-nominal attitude. To determine passive stability, engineers simulated a worst-case heat load scenario of an ascent abort, a re-entry from LEO, and a lunar return. For each of the scenarios the vehicle was not permitted to subject crew to loads exceeding the maximum allowable limits as set forth in NASA-STD-3000. After the simulations had been conducted it was determined that both vehicles were highly monostable and both were able to recover from off-nominal attitudes, although, all factors considered, the blunt-body design still had the performance edge.

Ultimately the blunt-body design won the trade studies thanks not only to its acceptable ascent and entry ballistic abort load levels but also because the design is a proven and familiar design permitting optimum crew seating, easier LV integration, and requiring fewer significant design challenges than the biconic. Since one of the prime drivers for shape selection was time, the blunt body had a clear advantage over the biconic.

THE DESIGN PROCESS

Although most spacecraft designers prefer to consider novel and innovative concepts, the reality is that the biggest constraints when designing a spacecraft are budget and timeline, neither of which favors unconventional technology. In the case of the Constellation Program, both the timeline and budget are restricted, which means designers must employ mature technology.

One of the first design process decisions is whether to build a spacecraft using separate elements to perform each function or to combine those functions in one integrated unit as, exemplified by the Space Shuttle system, the most highly integrated vehicle ever built. Although dividing functions is usually helpful when it comes to repair issues and often simplifies development, an integrated unit can reduce overall mass and volume, so program managers, mission planners, and engineers consider carefully the advantages and disadvantages of each process.

Design considerations step by step

Launch considerations

Launching the vehicle requires the design team to consider whether to divide the vehicle into multiple stages, which requires discarding items such as propellant tanks and other elements when no longer needed, or design a fully reusable vehicle that requires no elements to be spent. In the case of Orion, engineers decided to use a modular approach and include a SM in addition to the CEV.

Crew accommodation

The crew accommodations require the team to consider factors such as how much redundancy to build into the vehicle, how much habitable volume to provide, life support system design, and whether to provide just one or more crew compartments for the journey to and from the Moon.

Propulsion

Next on the list of considerations is the type of propulsion system. Liquid hydrogen and oxygen are very efficient but are difficult to store for long periods, whereas hypergolic fuels are less efficient but more reliable. It is possible the vehicle will have more than one means of propulsion, which means designers also have to decide if different propulsion types will use different propellant combinations.

Aerodynamics

The near-vacuum of space does not require a vehicle to be aerodynamic but this feature becomes extremely important when entering Earth's atmosphere not only because an aerodynamic vehicle increases stability but also because an aerodynamic shape reduces the thermal load generated on the vehicle surfaces. Regardless of which shape is chosen, one of the most important criteria engineers must consider is the lift-to-drag ratio (L/D), which describes the vehicle's ability to maneuver through the atmosphere. The most popular spacecraft shape is the blunt-nose capsule design of the Apollo and Soyuz vehicles.

Mode of landing

Mode of landing is largely determined by the vehicle's shape. A blunt capsule such as the Soyuz, for example, could never generate sufficient lift to land on a runway, although it could deploy a parafoil and glide to a somewhat controlled landing. When choosing the mode of landing, designers consider four factors. The first of these is whether the landing will be propulsive, by parachute, parafoil, wing, or lifting body. They then determine whether the descent will be vertical or gliding and how much maneuverability the vehicle will have during the descent. Finally, landing sites are considered, which in turn determines the options for attenuating the impact.

Further development

A "quantitative and qualitative assessment of design criteria" is probably an appropriate way of describing how design options are selected for further development. As some of these criteria may be more important than others, criterion-weighting factors are assigned in order to provide a weighted score for each option. For example, the large-scale ablative Thermal Protection System (TPS) that will be used on Orion is heavily weighted due to Orion's tight development schedule and the fact that a TPS of the size required by Orion has never been proven and therefore must be developed. Similarly, Orion's LAS is also heavily weighted due to the requirement to test new subsystems and the potential for design revisions that may lead to increased program costs and delays to the flight test schedule.

ORION VARIANTS

The ESAS was chartered to define and design Orion. Before beginning the design process the ESAS drafted ground rules and assumptions (GR&As) which included basic design requirements such as ensuring Orion would be capable of supporting a crew of up to six and be pressurized at 14.7 psi. Other design assumptions included the inclusion of a docking system, automated rendezvous and docking (AR&D) capability, two crewed flights per year, and three uncrewed, pressurized cargo flights per year. To cater for these various mission profiles it was necessary to design a number of Orion variants. Once the GR&As had been defined the study commenced and detailed subsystem definitions were developed and vehicle layouts completed for the various Orion designs. To keep track of Orion's various configurations, the ESAS study group assigned each design a block number based on the functionality of the vehicle.

Block 1A and 1B

The Block 1A (Table 4.6) was designed to have a total mass of 22,900 kg, including three crew, 400 kg of cargo, 8,300 kg of propellant, and 1,544 m/s of delta-V capability. It was envisaged that the Block 1A would be used to rotate between three and six astronauts and cargo to the ISS. The Block 1B version would be a replica of the Block 1A configuration but with all crew-related equipment removed. The Block 1B's primary mission would be to deliver 3,500 kg of pressurized cargo to the ISS and return an equivalent mass to Earth.

Cargo Delivery Vehicle

The Cargo Delivery Vehicle (CDV) (Table 4.6) was designed to replace the CM with an unpressurized cargo container capable of ferrying up to 6,000 kg of cargo before performing a de-orbit burn and being discarded in the ocean.

Table 4.6. Vehicle configuration summary.

	Block 1A ISS crew	Block 1B ISS pressurized cargo	CDV ISS unpressurized cargo	Block 2 Lunar crew	Block 3 Mars crew[a]
Crew size	3	0	0	4	6
LAS[b] required	4,218	None	None	4,218	4,218
Cargo capability[c]	400	3,500	6,000	Minimal	Minimal
CM (kg)	9,342	11,381	12,200	9,506	TBD
SM (kg)	13,558	11,519	6,912	13,647	TBD
Service Propulsion System delta-V (m/s)	1,544	1,098	330	1,724	TBD
EOR–LOR[d] 5.5 m total mass (kg)	22,900	22,900	19,112	23,153	TBD

[a] No detailed design details defined for Block 3 and no mass estimates calculated.
[b] Launch Abort System.
[c] Cargo capability equals total capability of the vehicle, including flight support equipment (FSE) and support structure.
[d] Earth orbit rendezvous–lunar orbit rendezvous.

Baseline Block 2

The Block 2 configuration (Table 4.6) was designated the baseline design. In this scenario a translunar injection (TLI) Earth Departure Stage (EDS) and an unmanned Lunar Surface Access Module (LSAM) would be launched by a Shuttle-derived heavy-lift Cargo Launch Vehicle (CaLV) and parked in an Earth parking orbit. Within 30 days, the Block 2 CEV would be launched using a Crew Launch Vehicle (CLV) and the CEV would rendezvous and dock with the LSAM and TLI EDS. Once joined, the spacecraft would be inserted into a translunar trajectory by the TLI stage, which would then be discarded. Lunar orbit would be achieved by a braking maneuver performed by the LSAM, after which the four crewmembers would transfer to the LSAM and descend to the lunar surface leaving the CEV in a quiescent mode in lunar orbit. Upon completion of the lunar surface mission, the crew would ascend in the LSAM ascent stage and rendezvous and dock with the CEV. After discarding the LSAM, the CEV would perform an engine burn placing it in a trans-Earth injection trajectory. Finally, on reaching Earth and aligned with the exact re-entry angle and position, the CM would detach from the SM and the CEV would descend under parachutes for a ground landing cushioned by airbags (Figure 4.4).

Figure 4.4.
Artist's rendering
of Orion landing
with airbags
deployed. Image
courtesy: NASA.

Block 3

Block 3 (Table 4.6) was not a part of the lunar mission architecture as it was designed for the reference Mars mission, which was also being considered in the ESAS study. The Block 3 configuration will transfer six crewmembers between Earth and a Mars Transit Vehicle (MTV) at the beginning and end of the Mars mission. The Mars mission architecture will require the CEV to be inserted into LEO before maneuvering and docking with the MTV in a higher orbit, after which the CEV powers down and remains in a quiescent mode for the 30-month Mars mission duration. On its return from Mars, the crew enters the CEV, undocks from the MTV, and uses the SM to make an onboard-targeted, ground-validated burn to aim for the critical entry corridor, after which the SM separates from the CM.

Design evolution of Orion

The Block 2 configuration of Orion evolved through four design cycles, beginning with a Cycle 1 baseline Apollo design 5 m in diameter. This iteration provided a pressurized volume of $22.3\,m^3$ for a crew of six, although this volume was barely sufficient once all equipment and stowage had been considered. Included in this design were $5\,g/cm^2$ of supplemental radiation protection.

Because of the restricted habitable volume provided by the Cycle 1 Orion, the Cycle 2 design cycle considered a larger vehicle 5.5 m in diameter, providing a pressurized volume of $39\,m^3$. This design also resulted in a net habitable volume of $19.4\,m^3$, an important increase since it would be necessary for astronauts to be able to stand up and don/doff EVA suits.

Table 4.7. Crew Exploration Vehicle design cycle evolution.

Configuration	Diameter (m)	Sidewall angle OML volume (m³)	OML volume (m³)	Pressurized volume (m³)
Apollo	3.9	32.5	15.8	10.4
Cycle 1	5.0	30.0	36.5	22.3
Cycle 2	5.5	25.0	56.7	39.0
Cycle 3	5.5	20.0	63.5	39.5
Cycle 4	5.5	32.5	45.9	30.6

Once the volume issue had been mostly resolved the third design cycle focused upon increasing Orion's monostability on Earth entry, achieved by reducing the sidewall angles. Although this had the added benefit of further increasing the interior volume, it had the disadvantage of increasing Orion's mass and upsetting the correct center of gravity (CG), a problem remedied by reducing the thickness of supplemental radiation protection.

The final design cycle was conducted following the decision to change the mission architecture and to remove the lunar surface direct mission profile. The Cycle 4 Orion was designed for a dual-launch Earth orbit rendezvous–lunar orbit rendezvous (EOR–LOR) mode in which Orion performs a rendezvous with the EDS and LSAM in low Earth orbit (LEO). The configuration then remains in lunar orbit while the LSAM descends to the lunar surface. The end result was a vehicle that bore more than a marked resemblance to the Apollo Command Module, although Orion had almost three times the internal volume of its predecessor. Table 4.7 tracks Orion's evolution through the four design cycles and compares the module sizing with the original Apollo vehicle.

ORION SYSTEMS AND SUBSYSTEMS

This section describes the systems and subsystems of each of Orion's four elements and the requirements of each demanded by the CARD. Given the complexity of designing and developing a new manned spacecraft, most of the section is necessarily directed at the CEV.

Orion Crew Exploration Vehicle

Vehicle overview

The Constellation Program requires Orion to perform manned and unmanned mission profiles but, given the Program's primary goal of returning humans to the Moon, the focus of this section is primarily directed at the lunar variant.

Before describing Orion's systems, and subsystems of the lunar CEV variant, it is appropriate to review the mission profile demanded of the vehicle. Together with the SM, LSAM, and EDS, Orion will be used to ferry four astronauts from Earth to lunar orbit and return them to Earth. In addition to providing habitable volume and life support for the crew, Orion also features a docking capability and a means of transferring the crew to the LSAM, in addition to re-entry and landing capabilities. On return to Earth, a combination of parachutes and four Kevlar airbags are deployed for a land touchdown, although a water flotation system may also be deployed in case of a water landing. Upon recovery Orion is refurbished and made ready for its next mission.

Vehicle shape

Orion's shape was largely a consequence of the familiar triad of cost, speed of design, and safety, which resulted in the decision to use a blunt-body capsule. One of the main advantages of choosing the blunt-body design lay in the familiar aerodynamic design which, thanks to the experience of Apollo, resulted in ascent, entry, and abort level loads that were familiar to engineers and consequently resulted in less design time and reduced cost. However, the shape is about the only design aspect that the Orion shares with its illustrious predecessor as it provides a much larger habitable volume and incorporates the very latest in avionics and life support technology (Table 4.8).

Table 4.8. Orion Crew Exploration Vehicle.

Configuration summary			
Diameter	5 m	Total CM delta-V	164 ft/s
Pressurized volume	691.8 ft	RCS engine thrust	100 lbf
Habitable volume	361 ft^3	Lunar return payload	220 lb
Habitable volume per CM	90.3 ft^3	CM propellant	GO_2/GCH_4
Mass properties summary			
Dry mass			7,907 kg
Propellant mass			175 kg
Oxygen/nitrogen mass/water			128 kg
CM landing weight			7,352 kg
Gross lift-off weight			8,502 kg

Vehicle materials

Orion's pressure vessel structure uses aluminum-2024 honeycomb sandwich for the face sheets and aluminum-5052 for the honeycomb core, a combination of materials that allow the vehicle to withstand the 14.7 psia internal cabin pressure required for the ISS rotation missions. For the purposes of rendezvous and docking operations, Orion is fitted with five double-paned fused silica windows, two forward-facing, two side windows, and a fifth window located within the side ingress/egress hatch.

Orion's outer mold line (OML) is composed of graphite epoxy/bismaleimide (BMI) composite skin panels, whose structure provides the vehicle's aerodynamic shape and serves as the attachment structure for the TPS.

Vehicle thermal protection

Orion's Thermal Protection System (TPS) must have the capability of shielding the crew from the thermal environment encountered during ascent, ascent abort, on-orbit operations, and re-entry. One of the most important factors engineers consider when deciding which material to use is the TPS heating rate for different entry trajectories, a design aspect that is reflected in the placing and thickness of the material. The thermal material that is used for Orion's TPS was determined by trade studies based on performance, mass, and cost. Among the materials considered were carbon–carbon, carbon–phenolic, phenolic-impregnated carbon ablator (PICA), PhenCarb-28, Alumina Enhanced Thermal Barrier-8 (AETB-8), and advanced reusable surface insulation (ARSI).

After much consideration, the material chosen by the Boeing Company, who was awarded the TPS development contract, was PICA. PICA was first used on the Stardust interplanetary spacecraft launched February 7th, 1999 to study the composition of the Wild 2 comet. When Stardust successfully re-entered Earth's atmosphere on January 15th, 2006, it was traveling at 46,510 km/h, the fastest re-entry speed ever achieved by a human-made object. Although Orion will not be traveling as fast as the Stardust spacecraft, it will need to withstand in excess of five times more heat (5,000°F) than spacecraft returning from the ISS. To protect Orion from the high heat flux generated during re-entry, the most effective protection is an ablative heat shield. As Orion enters the atmosphere, the extreme heat will cause the heat shield material to chemically decompose, a process known as pyrolysis. As PICA is pyrolized, the material will begin to char, before melting and finally sublimating, creating a cool boundary layer to protect the vehicle.

In January 2008, Orion's prototype heat shield underwent testing at NASA's Kennedy Space Center (KSC). The prototype (Figure 4.5), referred to as a manufacturing demonstration unit (MDU), has the same dimensions as the actual heat shield that will protect Orion as it re-enters Earth's atmosphere on its return from the Moon. The MDU currently rests in Hangar N at Cape Canaveral Air Force Station where it is undergoing several months of nondestructive laser scan and X-ray evaluation testing, the results of which will be used to construct the actual heat shield. The heat shield itself will consist of almost two hundred PICA blocks that share the same

Figure 4.5. The development of Orion's heat shield is being conducted at NASA's arc jet facilities at Johnson Space Center under the auspices of the CEV Thermal Protection System Advanced Development Project. The photo shows an arc jet test at the NASA Ames Interaction Heating Facility. The glowing material is a test coupon after test completion. Image courtesy: NASA.

delicate characteristics as the tiles on the Space Shuttle. Unlike the Space Shuttle, however, Orion's ablative TPS is designed to be used only once.

Vehicle propulsion

Orion's propulsion system comprises a Reaction Control System (RCS) that includes a number of elements such as the RCS tanks, the RCS pressurization system, the primary RCS thrusters, and the back-up RCS thrusters and RCS tanks. The RCS enables the Orion to perform exoatmospheric maneuvers and to orient itself during atmospheric re-entry. It also provides astronauts with a means of counteracting induced spin and dampening induced pitch and yaw instabilities which may occur during the lunar return trajectory. In the event of a loss of power, the Orion has a fully independent back-up RCS.

The propellant for the RCS is a bipropellant system comprising gaseous oxygen (GOX) and liquid ethanol. The GOX mixture, which also feeds the life support system, is stored in four cylindrical graphite-composite Inconel tanks mounted at the base of the vehicle, whereas the liquid ethanol is stored in two similar tanks and is pressurized by means of a high-pressure gaseous helium (GHe) system. Together, the

GOX and ethanol mixtures feed the twelve 445 N thrusters arranged in pairs to thrust in the pitch, roll, and yaw directions (Figure 4.6, see color section).

Vehicle power

Three primary rechargeable lithium ion batteries, 28-volt direct current (VDC) electrical power buses, power control units (PCUs), and back-up batteries comprise Orion's power subsystem, providing primary electrical power and distribution, and energy storage.

The lithium ion batteries were chosen due to the high amount of energy they provide, good charge retention, volume, and low drain life. The three primary batteries are capable of providing 13.5 kW/h for Orion's 2.25-hour storage requirement, the time between SM separation and landing. If more energy is required there is a fourth battery, providing one level of redundancy capable of providing 500 W of 28 VDC power for 45 minutes.

The lithium ion batteries feed electrical power to Orion's power distribution system together with the two "Mickey Mouse" solar arrays mounted on the SM (Figure 4.9). The power generated by these two systems results in an arrangement capable of distributing 28 VDC to the vehicle via a power distribution system comprised of a complex arrangement of jumper cables, brackets, cable ties, and primary and secondary distribution cables and avionics wiring.

While the distribution buses distribute power from the power sources, it is the function of the PCUs to monitor and control the current flowing from the solar arrays and batteries using a complex arrangement of switches, current sensors, and bus interfaces that ensure power is regulated according to load requirement.

Vehicle communications

Command and control (C&C) over all of Orion's operations is provided by the avionics subsystem comprised of command, control, and data handling (CCDH), guidance and navigation, communications, and cabling and instrumentation.

The role of the CCDH is to process and display important spacecraft data to the astronauts on multifunction liquid crystal displays (LCDs) and control panels. A part of the CCDH system includes two sets of translational/rotational/throttle hand controllers that enable the crew to take manual control of the vehicle when required.

The equipment providing the crew with on-orbit vehicle attitude, vehicle guidance, and navigation processing information is supplied by the guidance and navigation system. At the heart of the system is a Global Positioning System (GPS)/Inertial Navigation System (INS) that works in conjunction with two star trackers, video guidance sensors, and two three-dimensional scanning laser detection and ranging (LADAR) units.

Orion's communications components include S-band/search and rescue satellite-aided tracking (SARSAT), ultrahigh frequency (UHF) communications, information storage units, and a high-rate Ka-band communications system.

Orion's avionics

Orion features a state-of-the-art flight control system comprised of three briefcase-sized Honeywell flight control modules. Two of the control computers could completely fail and the third will still be able to fly the vehicle. In the worst-case scenario of a complete power failure Orion carries an emergency system powered by independent batteries that will provide the crew with enough capability to bring the vehicle back safely. In common with fighter aircraft cockpits Orion's control systems rely heavily on sensor fusion, a type of automation that relieves the astronaut-pilot of being a sensor integrator and allows him instead to focus on the mission. Such a system makes sense, given that many astronaut-pilots came from advanced cockpits such as the F-15 and F-22. In Orion's cockpit, astronaut-pilots will be able to change displays as if it's a revolving panel thanks to four flat-screen displays, each about the size of a large desktop monitor. During ascent to orbit the displays will operate similarly to the screens in conventional airliners. One display will show an artificial horizon, another will display velocity, and a third will show altitude. The fourth display will show life support status and communications information. Once Orion reaches orbit the displays will change to readouts showing rendezvous and docking information such as the vehicle's flight path, range, and rate of closing to the ISS.

Environmental Control and Life Support System

Orion's Environmental Control and Life Support System (ECLSS) includes all the items necessary to sustain life and provide a habitable environment for the crew. In addition to the equipment for storage of nitrogen and oxygen, the ECLSS also ensures the atmosphere inside Orion remains contaminant-free and provides fire detection and suppression capability.

The nitrogen gas required to sustain four crewmembers for nearly two weeks is stored in cylindrical graphite-composite Inconel-718 line tanks, while the oxygen is stored in the four primary RCS oxygen tanks. Atmosphere regulation is provided by a combined Carbon Dioxide and Moisture Removal System (CMRS), which ensures carbon dioxide levels are regulated, and an ambient temperature catalytic oxidation (ATCO) system for contaminant control. Fire detection and suppression capability consists of standard spacecraft smoke detectors and a fixed halon fire suppression system, identical to the ones used on the ISS. Other ECLSS components include cabin fans, air ducting, and humidity condensate separators, most of which are based on existing Shuttle technology or ISS systems.

Potable water is stored in four spherical metal bellows tanks, each capable of holding 53 kg of water. Once again, these tanks are similar to the ones installed on the Shuttle. In the event of a contingency EVA, which would require all crewmembers to don EVA suits and the cabin to be pressurized, the ECLSS includes the necessary umbilicals and ancillary equipment.

Active Thermal Control System

Orion's hardware will give off a lot of heat during operation. Since the behavior of fluids and the process of heat transfer are very different in microgravity, spacecraft engineers have had to design special equipment to radiate Orion's excess heat into space. To ensure a comfortable environment for the crew, NASA's Glenn Research Center (GRC), in partnership with the Jet Propulsion Laboratory (JPL) and Goddard Space Flight Center (GSFC), has developed the Active Thermal Control System (ATCS). The ATCS provides a temperature control capability for the vehicle consisting of a propylene glycol/water fluid loop with a radiator and fluid evaporator system. The fluid loop works as a heat rejection system by using cold plates for collecting waste heat from the equipment, while a cabin heat exchanger regulates atmosphere temperature.

To deal with high heat loads, the ATCS includes a dual-fluid evaporator system that works by boiling expendable water or freon-R-134A in an evaporator. This cools the heat rejection loop fluid which is circulated through the walls of the evaporator, which in turn causes vapor to be generated and then vented. The reason for a dual-fluid system is because water does not boil at the ATCS fluid loop temperatures and pressures at 30,000 m altitude or less, which means that from the ground to an altitude of 30,000 m freon-R-134A is used.

Crew living area

Although Orion provides a larger habitable volume than the Apollo vehicle, crew accommodations are fairly snug (Figure 4.7). Many of the facilities such as the galley are identical to those onboard the Shuttle, so lunar-bound astronauts will still be eating freeze-dried and irradiated pre-packaged foods. There is little in the way of improvements in the toilet department which is a passive Mir-style design, comprising a privacy curtain, contingency waste collection bags and what NASA euphemistically refers to as a suitable user interface! Given the spartan toilet facilities it is probable that many astronauts will adopt appropriate nutritional intervention strategies prior to launch to limit the number of toilet visits!

Non-propellant

Oxygen, nitrogen, potable water, FES water, and freon constitute Orion's non-propellant components. Oxygen is used for breathing by the crew but also for contingency EVA consumption and in the event of rapid or explosive decompression which would require full-cabin repressurization. Mission planners have budgeted for 64 kg of oxygen for the round trip to the Moon, an amount that includes cabin leak rates and contingency supplies. Nitrogen is used for the cabin atmosphere and also for waste management and the Carbon Dioxide Regeneration System. The mass budget of nitrogen for the trip is 32 kg.

Potable water is required not just for drinking but also for contingency EVA use and Orion's water evaporator system. Mission planners have estimated that each crewmember will use 3.5 kg of water per day for food preparation, hygiene, and drinking.

Figure 4.7. Cut-away of Orion showing crew compartment. Image courtesy: NASA.

Parachute and Landing System

The Constellation baseline for Orion's return to Earth was a land landing. The main reason for initially deciding to land on solid ground was to reduce the damage to the vehicle's TPS. However, in the summer of 2006, after examining the safety and cost aspects of land vs. water landings, the decision was taken to change the baseline to a water landing, which will occur near San Clemente Island, northwest of San Diego. However, in the event of an off-nominal/abort contingency event Orion will still be capable of a land landing thanks to a Wraparound Partial Airbag System comprising four cylindrical airbags located on the "toe" of the capsule.

Orion's Parachute System is packed between the vehicle's pressure vessel and OML, which is near the docking mechanism. Consisting of two 11 m diameter drogue parachutes (Figure 4.8) and three round primary parachutes (Figure 4.9), each with a diameter of 34 m, the Parachute System ensures a nominal landing speed of 8 m/s with all three parachutes deployed and a landing speed of 8.5 m/s with one failed parachute. The system is automated much like the automatic opening device (AOD) used to open civilian parachutes, which deploy canopies using a sensor that detects a dynamic pressure. In the case of Orion's drogue system, the dynamic pressure is dialed in to 7,000 m altitude and a 400 km/h sink rate. The drogue chutes slow the sink rate to 200 km/h which is achieved at an altitude of 3,300 m, which is the deployment altitude for the primary system.

In the event of a land landing, Orion is cushioned by four inflatable Kevlar

Figure 4.8. Artist's rendering of Orion's drogue chutes deployed. Image courtesy: NASA.

airbags. As the vehicle descends the airbags are deployed out of the lower conical backshell. Two panels jettison, permitting the airbags to inflate and wrap around the low hanging corner of the heat shield to provide energy attenuation upon landing. Once Orion has landed, the airbags vent at a specific pressure to facilitate a controlled collapse rate.

Figure 4.9. Artist's rendering of Orion's three parachutes deployed. Image courtesy: NASA.

Figure 4.10. Langley's Landing and Impact Research Facility is used to evaluate landing and attenuating systems like airbags, retrorockets, and passive energy–absorbing struts for Orion. The gantry is 73 meters high, and is a steel truss structure that has sometimes been described as a Lego erector set on steroids. It's a series of three A-frames, consisting of a canted leg 73 meters high, a horizontal component, and another leg on the opposite side canted at the same degree down to the ground. At the top of the gantry, crosswalks connect all three A-frames, and at one end there is a bridge. The bridge has a winch on it used to pull the test article back to the drop height to release it for the test. Image courtesy: NASA.

Experimental tests are ongoing at the NASA Langley Landing and Impact Research (LandIR) Facility, in Hampton, Virginia (Figure 4.10), to evaluate the effectiveness of different airbag designs and their ability to absorb the shock of landing. The LandIR facility was previously used by Apollo astronauts who practiced descending to the lunar surface in a simulated lunar module. The facility consists of a 73 m high gantry structure and crosswalks that connect three A-frames supporting a bridge from which DTAs are released. Forty years after Apollo, LandIR has been configured to drop DTAs fitted with and without airbags from heights as great as 10 m onto surfaces covered with sand and/or clay, which simulate the Orion landing sites such as the dry lakebeds in California. The test campaign utilizes two DTAs, one a half-scale metal shell and the other a geometric full-scale fitted with shock-absorbing airbags.

The half-scale models were designed for model performance studies, soil impact characterization, and training test personnel and were not fitted with an airbag attenuation system. The full-scale airbag drop model is fitted with either three or four airbags by virtue of it weighing less than half the weight of the actual Orion module, which will be outfitted with six airbags. After 20 test drops conducted between December 2006 and June 2007, the current conceptual airbag design comprises an internal airbag, an orifice assembly, and an outer, venting airbag.

The design of the conceptual airbag model was derived from single-camera, single-view photogrammetric analysis of the test drops. This type of analysis involved positioning photogrammetry targets on the airbags and then filming the test drops using high-speed (1,000 frames per second) video cameras placed at various angles to the drop trajectory. To accurately determine the effects of initial impact, attitude changes, and first bounces, testers used frame-by-frame tracking, progressing sequentially forward through the video frames, making a note of the position of each photogrammetric target. This information was then processed by software to determine the exact time of impact, horizontal and vertical trajectory angles, out-of-plane motions, and displacement and velocity changes as the airbag impacted the ground. The tests began with vertical drops from low heights and no induced pitch or yaw angles and progressed to higher pendulum-style drops that featured considerable pitch and yaw angles.

Orion weight issues

Throughout its design history, Orion has been plagued with weight issues. One of the headaches facing mission planners and spacecraft engineers is the issue of mass and how to reduce it to a minimum. One of the targets for mass reduction is non-cargo, which usually consists of personnel, personnel provisions, and residual propellant. Since Orion is designed with the assumption of a 95th-percentile crewmember weighing 100 kg, one of the easiest ways to reduce weight is personnel, since, so far in NASA astronaut history, no crewmember has weighed this much. Another target for weight reduction is personnel provisions for the crew which include recreational equipment such as crew preference items and health care supplies. For a lunar trip crewmembers will need to pack light as they will be permitted to take only 5 kg of recreational equipment with them. One reason for the restrictions of recreational equipment is that personnel provisions such as crew health care equipment also weigh a fair amount. In addition to the dental and surgical supplies, mission planners must also find room for personal hygiene kits, crew clothing, housekeeping supplies, and operational supplies such as zero-g restraints, emergency egress kits for pad aborts, binoculars, crew survival kits, and a life raft that weighs 44 kg. Also included in the personnel provisions are the EVA suits and spares, which weigh 20 kg each, and the toolkit in case the crew need to effect repairs.

Unfortunately, Orion's design resulted in weight problems that could not be solved by simply asking astronauts to pare down their preference items. To save weight, engineers examined the possibility of paring down the vehicle's land landing system. Since water landing was a perfectly viable landing mode it just didn't seem efficient to carry 680 kilograms of landing bags to the Moon and back. Other engineers argued that using water landing as a baseline would compromise the vehicle's integrity due to saltwater exposure thereby reducing Orion's reusability.

Table 4.9. Service Module.

Configuration summary			
Structural configuration	Three rings/six longerons	RCS thruster thrust	100 lbf
Propulsion configuration	2 × 2 serial feed	Solar array area	388 ft²
SM propellant	MMH/N₂O₄	Solar array power	9.15 kW
Total SM delta-V	6,086 ft/s	Radiator area	334 ft²
Main engine thrust	7,500 lbf	Thermal dispersion	6.3 kW

Ultimately, the requirement to save significant weight resulted in a change to the baseline from a land landing to water landing.

Service Module

Vehicle

The function of the SM (Table 4.9) is to provide maneuvering capability, power generation, and heat rejection for Orion. The vehicle (Figure 4.11) features a 66.7 kN Service Propulsion System and a RCS comprising twenty-four 445 N thrusters enabling it to conduct rendezvous and docking with the LSAM in Earth orbit. While in flight the SM will also cover and protect Orion's heat shield.

Power

Orion's power during Earth orbit insertion (EOI) to Orion–SM separation prior to entry is provided by two deployable, single-axis gimbaling solar arrays that use state-of-the-art junction photovoltaic (PV) cells. The reason for choosing solar arrays is the requirement of Orion to remain unoccupied in lunar orbit for up to 180 days, a period of dormancy considered too risky to rely on fuel cells or other power generation options.

A Power Management and Distribution (PMAD) System ensures that Orion receives adequate power, allowing for factors such as solar array degradation and losses incurred by the arrays not pointing at the Sun at the correct angle. Power is distributed as 28 VDC to SM loads, using a system similar to the one used to distribute power to Orion.

Construction

The SM is a semimonocoque unpressurized structure similar in design to the Apollo SM. The structure provides attachment for Orion's avionics, propulsions system components, and an interface for mating to the LM. Constructed of graphite

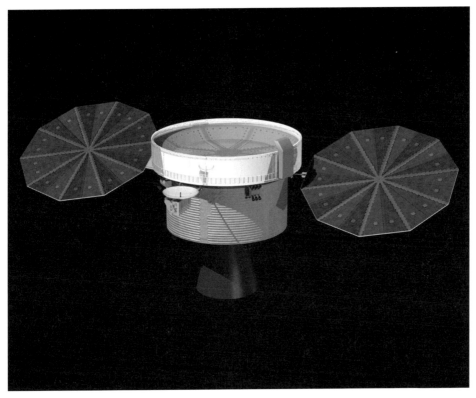

Figure 4.11. Orion's Service Module provides the astronauts with power thanks to two deployable, single-axis gimbaling solar arrays. Image courtesy: NASA.

epoxy/BMI composites and aluminum alloys, the SM also contains insulation blankets for thermal control.

Propulsion

A pressure-fed integrated Service Propulsion System/RCS using LOX and liquid methane (LCH$_4$) comprises the SM propulsion. LCH$_4$ was selected due to its high I_{SP} qualities, nontoxicity, commonality with the LSAM, and its suitability for use for *in situ* resource utilization (ISRU). Whereas the propulsion system is used for major translational maneuvers and vehicle attitude control, minor translational maneuvers are performed by 24 oxygen/methane pressure-fed RCS thrusters. The propellants for the Service Propulsion System and RCS are stored in four cylindrical tanks constructed with Al–Li-2090 liners and graphite epoxy over wrappings. Oxygen and methane are stored in tanks constructed of variable-density multilayer insulation (MLI). Pressurization for the propellant tanks is provided by GHe stored in Inconel-718 lined graphite epoxy-composite tanks. As propellant is used the pressure in the tanks is maintained by GHe being distributed to the oxygen/methane tanks.

Avionics

The SM features an avionics subsystem that performs similar functions to Orion's subsystem. The CCDH includes interface units that collect and transmit health and status data from other SM components, which is then transmitted to Orion's CCDH system.

Active Thermal Control System

The SM's ATCS is comprised of a single-loop propylene glycol fluid loop and a radiator mounted in Orion, except for the radiator panels which are fixed on the SM body.

Miscellaneous components

Since Orion and the SM are mated, the SM includes a vehicle interface and umbilical lines so the two vehicles can share power and, to enable separation, the SM also carries a pyrotechnic separation mechanism.

Non-cargo

The primary non-cargo item of the SM is residual propellant which consists of oxygen and methane trapped in the propulsion tanks following nominal delta-V maneuvers. The oxygen and methane are stored as cryogenic fluids and stored passively, meaning there is no attempt to stop propellant vaporization. Therefore, as heat seeps into the propellant tanks, the remaining cryogenic propellant slowly turns to gas and all that remains is a means to vent it, which is performed automatically by the pressurization system.

Propellant and delta-V maneuvers

The SM's propellant is calculated for four delta-V maneuvers, the first of which is the rendezvous and docking with the LSAM in LEO. In this maneuver Orion is placed into a 55 × 185 km elliptical orbit by the Upper Stage (US) of the LV, while the LSAM and EDS follow a 296 km circular orbit. In order for Orion to rendezvous with the LSAM, a delta-V must be performed.

The second maneuver is comprised of two "station-keeping" maneuvers, each of which is performed in LLO while the crew is on the lunar surface. The first of the station-keeping maneuvers is simply maintaining LLO, requiring minimal delta-V, whereas the second maneuver is a contingency plane change in the event the crew must abort from the surface.

The trans-Earth injection (TEI) from LLO is Orion's third maneuver, which requires a high delta-V. Also, there must be sufficient propellant in case of a worst-case-scenario anytime return from a polar orbit, an event that would necessitate a very high delta-V due to the requirement to effect a major plane change needed to align the spacecraft's velocity with the departure vector.

Table 4.10. Spacecraft Adapter.

Length	3.31 m
Basic diameter	5.03 m
Maximum diameter	5.50 m
Mass	581 kg

The final maneuver is the lowest delta-V that requires only a 10 m/s course correction burn using an RCS, although the burn required to discard the SM following separation from Orion may be considered a fifth maneuver.

Spacecraft adapter

The SA (Table 4.10) is a simple aluminum structure covered in white silicone thermal control coating. At the base of the structure is a field joint that attaches to the FS and vent holes that equalize pressure during ascent.

Launch Abort System

The LAS is designed to pull Orion away from the thrusting LV First Stage in the event of an off-nominal situation (Figure 4.12, see color section). Compared with many other mission elements, the LAS is a relatively simple design sharing many features with the Apollo Launch Escape System (LES). The system (Table 4.11) incorporates an active tractor design that utilizes a canard section below the attitude control motor element. Below the canard section are four jettison motors which sit atop the systems interstage. Below the interstage is the abort motor element, comprising four exposed, reverse flow nozzles. Attached to the aft end of the abort motor element is the adapter cone which in turn is attached to the boost protective cover (BPC). Following Second-Stage ignition, the LAS is discarded, after which any contingency for abort is provided by the SM Propulsion System.

ORION ABORT MODES

[CA0466-PO]   The CEV shall perform aborts from the time the CEV abort system is armed on the launch pad until the mission destination is reached. *Rationale:* Abort at any time is part of NPR 8705.2, Human Rating Requirements for Space Systems, as well as the program policy on crew safety. This CEV requirement will cover all of the flight phases from abort system arming on the launch pad through docking with the ISS or LSAM landing. The CEV must be capable of supporting an LSAM descent abort and subsequent redocking. After reaching the destination, all other scenarios are covered by the return capabilities.

Table 4.11. Launch Abort System summary.[a]

Configuration summary					
Abort motor		Attitude control motor		Jettison motor	
No. of nozzles	4	No. of nozzles	8	No. of nozzles	4
Nozzle cant angle	300	Nozzle cant angle	900	Nozzle cant angle	350
I_{SP} (sea level)	250 s	I_{SP} (vacuum)	227 s	I_{SP} (vacuum)	221 s
Thrust (total in vehicle axis)	506,408 lb	Thrust (per nozzle)	2,500 lb	Thrust (per nozzle)	9,668 lb
Burn time	2.0 s	Burn time	20 s	Burn time	1.5 s
Performance summary					
System thrust-to-weight (T/W) ratio				15 for 2.0 s	
Pad abort altitude				>4,000 ft	
Pad abort downrange				>3,500 ft	
Typical separation distance after 3 seconds				>1,500 ft	
Max dynamic pressure abort separation distance after 3 seconds				>600 ft	
LAS lift-off gross weight				12,977 lbm	
Mass properties summary					
Dry mass				8,148 lb	
Gross lift-off weight				13,616 lb	
Propellant				5,468 lb	

[a] Adapted from Orbital Sciences Corporation.

The CARD states there shall be no period during a mission in which a survivable abort mode is not available to the crew. The CARD also requires that no abort mode shall land Orion in the North Atlantic Ocean more than 250 km from St. John's, Newfoundland or Shannon, Ireland, an area also known as the North Atlantic Downrange Abort Exclusion Zone (DAEZ). These stipulations have required engineers to perform crucial analysis of propellant loading since Orion's modes of abort are driven by the relatively low lift capability of Ares I. Engineers must also consider the delta-V capabilities that may allow a crew to abort-to-orbit (ATO) instead of a landing in the North Atlantic or possibly in the continental United States (CONUS).

To determine abort capabilities, engineers assessed two propellant loadings, one which maximized the earliest targeted abort landing (max-TAL) and one which maximized the earliest ATO capability (max-ATO). For each of the propellant loadings, engineers calculated five abort thrust magnitudes (7,500 lbf, 8,300 lbf, 10,000 lbf, 10,800 lbf, and 10,980 lbf), each magnitude representing various combinations of main engine and auxiliary engine thrust levels. For the purposes of calculation, the specific impulse of 323 seconds remained the same. Next, engineers identified three abort modes concerned with system design impacts of the North Atlantic DAEZ requirement, the first of which was designated the untargeted abort splashdown (UAS).

In the event of an UAS (Table 4.12), which will result in a water recovery off the Atlantic coast of either the United States or Canada, the SM's RCS fires to separate Orion from the Ares US, after which astronauts fire Orion's thrusters to control and guide the vehicle's bank angle during re-entry, being very careful to avoid acceleration levels that might subject them to excessive g-loads.

The next abort mode engineers analyzed was a targeted abort landing (TAL), which requires the crew to first use the SM RCS to separate from the Ares I US and then fire the OME to boost the landing area to within 250 km of Shannon, Ireland. This type of abort (Table 4.12) is available once the SM Propulsion System is capable of ensuring the required delta-V while maintaining a minimum altitude of at least 121,200 meters to avoid excessive heat on re-entry.

The ATO abort mode uses the SM RCS to separate Orion from the Ares I US and uses the OME to increase the apogee altitude to approximately 160 km, after which Orion coasts to apogee and performs an insertion burn to ensure a stable orbit. The main reason for increasing the altitude is the constraint imposed by thermal heating

Of the three abort modes, the most desirable is the ATO mode since this allows the possibility of continuing a nominal mission or at least landing the crew within CONUS, which ensures safer recovery operations.

The three abort modes described were subjected to analysis using the Optimal Trajectories via Implicit Simulation (OTIS) software program [4] which breaks down the trajectory sequences into phases, which are then used to model specific parameters of each abort mode. For example, one phase common to each abort mode is the "abort initiation phase". By using OTIS, engineers determine factors such as the earliest abort time possible, given constraints such as vehicle configuration and environment. Once these constraints are considered, OTIS considers other factors that may affect the abort, such as altitude, latitude and longitude, velocity, relative flight path angle, and azimuth. In fact, OTIS can calculate dozens of variables for each phase of the trajectory sequence beginning with engine ignition until the abort initiation, thereby providing engineers with a precise modeling of possible reference trajectories.

The end result of using OTIS to analyze the three abort modes was to provide engineers with thrust-to-weight ratios, propellant loadings, and other factors that may affect the performance of Orion during each abort mode. It also provided engineers with clearly defined abort windows for given engine configurations and confirmed that abort mode overlaps were sufficient and in accordance with Constellation requirements.

Table 4.12. Orion abort modes.[a]

Abort mode	Phase	Description
Untargeted abort splashdown (UAS)	Abort initiation	• Abort is initiated at t_0 • Orion's state is assigned based on interpolated CLV at t_0 • Orion coasts to re-entry interface
	Re-entry	• CM separates from SM at re-entry interface • Initial pitch angle interpolated from trimmed aerodynamic database based on Mach number • CM re-enters atmosphere. Bank angle optimized so abort initiation may occur as early as possible while still landing within 250 km of St. John's International Airport
Targeted abort landing (TAL)	Abort initiation	• Abort is initiated at t_0 • Orion's state is assigned based on interpolated CLV at t_0
	Separation	• CM separates from SM and drifts for 15 s
	Main engine burn	• Main engine and auxiliary thrusters fired to boost downrange landing point into TAL recovery area, while maintaining altitude limitation of 121,200 m
	Re-entry interface	• Orion coasts to re-entry interface at 90,900 m
	Re-entry	• Orion re-enters atmosphere, deploys parachutes at 15,150 m, and lands near TAL recovery area
Abort to orbit (ATO)	Abort initiation	• Abort is initiated at t_0 • Orion's state is assigned based on interpolated CLV at t_0
	Separation	• CM separates from SM and drifts for 15 s
	Main engine burn	• Main engine and auxiliary thrusters fired to boost Orion's apogee altitude to 160 km
	Coast to apogee	• Orion coasts to almost apogee
	Circularization	• Orion circularizes using main engine

[a] Adapted from Falck, R.D.; Gefert, L.P. *Crew Exploration Vehicle Ascent Abort Trajectory Analysis and Optimization*, NASA/TM-2007-214996. Glenn Research Center, Cleveland, OH.

RISK ASSESSMENT

In July 2006, a Systems and Software Consortium (SSC) Risk Assessment Team generated a risk profile for Orion using a modified Delphi technique developed by the RAND Corporation in the 1940s. The Delphi technique, named after the ancient Greek oracle at Delphi that was believed to predict the future, is characterized by anonymity, controlled feedback, and statistical response. To generate the risk assessment, the SSC team followed a three-step process starting by sending a questionnaire to a panel of experts. Responses from the experts were then iterated and risks categorized, after which a face-to-face meeting with the experts was facilitated using the Electronic Meeting System (EMS).

Some of the experts questioned were NASA personnel such as Associate Administrator of Safety Bryan O'Connor and NASA Risk Analyst Bill Cirillo. Other experts were non-NASA personnel such as John Kara, Lead of the Advanced Launch Vehicle Program at Lockheed Martin, and ex-NASA veteran George Mueller, who works as Program Manager for Kistler Aerospace. The experts were sent a questionnaire which asked them to provide considered opinion regarding the risks for Orion's Navigation System, ECLSS, Propulsion System, and other critical systems. Following iterations of questions and responses among the experts and after conducting the face-to-face meeting via EMS, the SSC Risk Assessment Team defined nine categories of risks, including risks assigned to systems complexity, systems architecture, re-entry failure modes, and integration, verification, and validation issues. Next, the SSC team formulated 14 questions that addressed each category of risks. For example, Question 6 asked what the risks were due to parts, materials, and components selection, while Question 8 asked what risks were associated with the launch. A grid was then generated, tabulating all risks for each question and category, a process that resulted in approximately 600 risks being identified, many of which fell under the categories of system design and development (>150); integration, verification, and validation (>100); and programmatic engineering issues (>90). Some of the major risks to the crew included inadequate TPS and radiation protection, inadequate AR&D, a lack of LAS development, insufficient Orion systems reliability, issues with NASA decision-making culture, and failure of the solid rockets.

Since the SSC assessment, many of the risks to Orion have been resolved thanks to a cascade of changes constituting the evolution of the vehicle. Equally, the mass-to-orbit problems of the Orion design are slowly evolving away thanks to a redesign of the Service Module which reduced the vehicle's mass by 450 kg. Other weight-saving possibilities continue to be studied such as the module's radiators and changing the TPS material. To give engineers more time to work out some of the weight-saving decisions the Orion program slipped its PDR from September to November 2008.

As the centerpiece of the United States' new space exploration program, Orion will have a demanding career. While the vehicle's appearance may seem humble in comparison with the Space Shuttle, the endeavor Orion represents is anything but. Although the vehicle's first mission will be to deliver six astronauts to the ISS and remain docked there as a lifeboat for six months, the focus of Orion's design is to fly a crew of four to the Moon, a mission it will achieve thanks to the rigorous design and development process described here.

REFERENCES

[1] NASA. *Constellation Architecture Requirements Document*, NASA CxP 70000. Baseline. NASA, Washington, D.C. (December 21, 2006).

[2] Petro, A. Transfer, entry, landing, and ascent vehicles. In: W.J. Larson and L.K. Pranke (eds.), *Human Spaceflight: Mission Analysis and Design*, pp. 392–393. McGraw-Hill, Columbus, OH (2000).

[3] NASA. *Man–Systems Integration Standards*, NASA-STD-3000. NASA Johnson Space Center, Houston, TX (1995).

[4] Riehl, J.P.; Paris, S.W.; Sjauw, W.K. *Comparisons of Implicit Integration Methods for Solving Aerospace Trajectory Optimization Problems*, AIAA 2006-6033. American Institute of Aeronautics and Astronautics, Washington, D.C. (August 21–24, 2006).

5

Lunar outpost

When NASA astronauts return to the Moon in 2020, the first missions will be no longer than seven days. To sustain extended-duration lunar surface missions, a lunar infrastructure must be capable of supporting the basic functions of surface operations such as construction and maintenance procedures, surface transportation, crew support activities, and other support operations such as communication and navigation. This chapter describes the facilities and equipment that will constitute the lunar outpost and discusses the design details of the primary elements required to establish a permanent human settlement on the Moon.

LUNAR ENVIRONMENT

Engineers are currently developing concepts for habitation modules that will provide the necessary protection and support for crewmembers embarking upon long-duration missions on the lunar surface. In the process of designing and developing these modules, engineers must consider the challenges presented by the inhospitable lunar environment. Given the harshness of the lunar surface and the fact that many of its characteristics do not exist on Earth, it is appropriate to begin with an overview of what constitutes the location of Earth's first off-world outpost.

Geography

The highlands, maria, and craters comprise the Moon's primary geographical features. The highlands, the most ancient parts of the lunar surface, are distinctive due to their very rough features, a consequence of millions of years of meteorite bombardment. Also referred to as the *terrae*, the lunar highlands make up 66% of the Moon's area visible from Earth and comprise 83% of the total lunar area. In contrast, the maria, or lowlands, occupy 17% of the Moon's surface and are its smoothest features since they were formed by lava flows [9].

Geology

Lunar rocks are grouped into basaltic volcanic rocks, pristine rocks, breccias, and impact melts. The basaltic volcanic rocks are found in the maria and are rich in iron and titanium, most of which is found in the mineral *ilmenite*. Pristine rocks are found in the highlands and contain potassium, phosphorus, and pyroxene, whereas breccias are found mostly in craters and are the product of micrometeoroid impacts that give rise to shattered rocks or impact melts.

In addition to the different classes of lunar rocks there is the lunar regolith which includes rock debris of all kinds, including volcanic ash. Although lunar regolith is often grouped together with lunar soil, the latter refers only to the sub-centimeter fraction of the lunar regolith. Regolith usually contains high quantities of iron, whereas lunar soil is composed mostly of agglutinates which contain *troilite*, a non-magnetic form of iron sulfide [9].

Gravity

Lunar gravity is only one-sixth of Earth's gravity so any lunar structure will, in gross terms, possess a six times greater structural weight bearing capacity on the Moon than on the Earth. For this reason and to maximize the utility concepts developed for lunar structural designs, it has been suggested that mass-based rather than weight-based criteria be used to design lunar structures [3].

Temperature

In conjunction with lunar day–night transitions, lunar surface temperatures may vary from 107°C to −153°C [17], a thermal variation caused by the amount of radiation absorbed by the Sun and lack of a significant lunar atmosphere (Table 5.1).

Atmosphere

The Moon is surrounded by a tenuous lunar atmosphere which is almost a hard vacuum. This means that engineers, rather than being concerned about wind loads, will need to focus on designing a habitat constructed of materials stable enough to tolerate conditions where air pressure does not exist.

Radiation

The lunar surface is subject to electromagnetic and ionizing radiation. Electromagnetic radiation will actually be helpful to lunar dwellers as it provides a solar power source. Ionizing radiation, however, consisting of protons, electrons, and heavy nuclei, has the potential to penetrate materials in depths ranging from a few microm-

Table 5.1. Estimated lunar surface temperature [9].

	Shadowed polar craters	Other polar areas	Front equatorial	Back equatorial	Limb equatorial	Typical mid-latitudes
Average (°C)	−233	−53	−19	−17	−18	−53 to −18
Monthly range (°C)	None	−263	−133	−133	−133	−163

eters to several meters and are capable of inflicting serious damage upon crewmembers living on the lunar surface [5].

The various types of ionizing radiation include galactic cosmic radiation (GCR), solar cosmic rays (SCRs), and solar wind. The latter type of ionizing radiation consists of an electrically neutral stream of ions and electrons emitted from the Sun at a velocity of between 300 km and 700 km per hour. As most of the solar wind particles are of low energy, the damage inflicted by them is limited. SCRs, however, contain heavy nuclei with high energies and have the potential to inflict significant damage. Perhaps the most damaging radiation type, however, is the radiation generated by GCRs [15]. Composed of high-energy high-nuclei particles, GCRs consist of intense, penetrating radiation that has the capability to kill any crewmember not adequately protected.

Meteoroid

A meteoroid penetrating a habitat or a spacesuit may result in either a rapid or explosive decompression of the habitat or suit and put crewmembers at risk of ebullism and/or severe decompression sickness (DCS). To protect crewmembers against such an event, NASA utilizes a means of predicting when meteoroid strikes may occur. The currently accepted model for predicting meteoroid strike incidence is contained in NASA Technical Memorandum 4527 [1], developed by the Marshall Space Flight Center (MSFC), which utilizes operational altitude, meteoroid diameter, and particle density and mass to predict meteoroid fluence. Based on the MSFC model, meteoroid fluence for the Moon varies between 1,088.42 meteoroid impacts/m^2/yr for meteoroids of 1×10^{-4} cm or greater in diameter to 1.61142×10^{-18} meteoroid impacts/m^2/yr for meteoroids of 700 cm or greater in diameter.

Lunar dust

"I am at the foot of the ladder. The LM footpads are only depressed in the surface about one or two inches, although the surface appears to be very, very fine-grained, as you get close to it, it's almost like a powder; down there, it's very fine. I'm going off the LM now. That's one small step for man. One giant leap for mankind. The surface is fine and powdery. I can pick it up loosely with my toe. It does adhere in fine layers like powdered charcoal to the sole and sides of my boots. I only go in a small fraction of an inch. Maybe an eighth of an inch, but I can see the footprints of my boots and the treads in the sandy particles."

Neil Armstrong's initial impression of lunar dust.

Apollo astronauts discovered that the lunar surface is covered with a layer of easily disturbed fine particles. Due to its highly abrasive nature, lunar dust has the potential to pose a serious risk not only to the health and wellbeing of crewmembers but also to equipment. Extensive evidence confirms the significant pulmonary and cardio-vascular effects of both acute and chronic exposure to lunar-like particulate matter air pollution smaller than 10 μm. Research also documents that chronic occupational exposure to dust is associated with progressive non-malignant pulmonary disorders and disease. Unsurprisingly, space medical doctors are concerned that lunar dust, due to its sharp and jagged surface, may inflict similar damaging effects upon resident astronauts [18].

Seismic activities

Due to the lack of plate tectonics, seismic activity on the Moon is very low. The release of seismic energy from the lunar interior is about seven orders of magnitude lower than that of Earth and so the Moon is considered seismically stable. Although the seismometers left behind on the lunar surface by the Apollo astronauts recorded seismic activity for nearly eight years, during that time no significant moonquake was recorded.

Outpost location

The site of NASA's outpost has yet to be decided, but the aforementioned surface characteristics will surely play a part in determining the optimal location. Based on current knowledge it is expected the outpost will be situated at one of the lunar poles. There are five reasons for choosing a polar site, the first being the abundance of sunlight, which alleviates concerns regarding energy storage, Second, the polar environment is relatively benign, characterized by temperatures that vary no more than 50°C year round, compared with the 250°C temperature variation at the lunar equator. Third, at the South Pole there is evidence of hydrogen, an important natural resource for developing propellant production. Fourth, the poles are among the most complex geological regions of the Moon, making them a natural target for scientific study. Lastly, landing equipment and scientific payloads at the South Pole compared with other locations requires less fuel and results in more cost-effective missions.

Presently, the leading contender as an outpost location is the Shackleton Crater, located near the South Pole. A feature roughly 19 kilometers in diameter, the crater is named after legendary Antarctic explorer, Ernest Shackleton. Despite the harsh terrain characterized by peaks as high as Mount McKinley and crater floors four times deeper than the Grand Canyon, the area is an explorer's paradise. Mission planners say the area is ideal for hosting a lunar outpost thanks to the area being illuminated by sunlight for a substantial part of the day, meaning solar panels could provide the outpost with a steady flow of electricity. The crater is believed to be 3.5 billion years old, which means it is not as rocky or steep as younger, fresher craters, a feature corroborated by recent radar data.

Lunar infrastructure

NASA's current plan for establishing a lunar outpost is based on landing up to three large habitats, using unmanned cargo flights, a strategy directed at getting the outpost up and running quickly. Such an approach means astronauts will have the opportunity to reap science rewards faster than if they have to build the outpost incrementally by hauling small habitat modules and hardware to the Moon on each flight.

Exactly what these habitats will look like is unknown as the first lunar flights are still more than a decade in the future. What is known is that the design NASA ultimately chooses will be constrained geometrically and by weight due to the shroud volume of Ares V and the shape of the Cargo Lunar Lander (CLL), which will transport the habitat to the lunar surface. Due to these constraints it was decided a single habitat could not be flown, and instead habitat elements would need to be flown separately.

Although NASA is currently evaluating concepts for habitation modules, to date no baseline lunar habitat or lunar outpost design exists. The objective of this section, therefore, is to describe the lunar habitat design process based on work conducted by Orbital Sciences Corporation's (OSC) Advanced Programs Group (APG) and to explain the challenges faced by mission planners and engineers.

Lunar Habitat concept

Following the announcement of the President's Vision for Space Exploration (VSE), OSC's APG developed a lunar exploration architecture under contract to NASA as a part of the Concept Exploration and Refinement (CE&R) program. The CE&R program was developed to advise NASA on feasible requirements pertaining to the future of U.S. space exploration. As a part of the lunar exploration architecture (described in Chapter 8) OSC developed the Lunar Habitat concept described here.

One of the first considerations OSC addressed was defining the optimum Lunar Module shape, by performing trade studies of different module geometries (Table 5.2). The trade studies assessed seven different configurations based on factors such as weight, habitable volume, and shell thickness (Table 5.2).

Due to the density of the metal used in the trade studies, radiation protection was limited and so this criterion was not considered. A manufacturing limitation placed on the evaluation was the assumption of the use of Al–Li-8091 material with a minimum-gauge thickness of 0.152 cm. Having decided upon the modules' design, the team then evaluated two cases for each geometry.

The "A" cases each had a uniform thickness stemming from the maximum stress location, whereas the "B" cases optimized thickness for different areas of the module,

Table 5.2. Habitat Module geometry analysis [4].[a]

Case	Shape	Total volume (m³)	Livable volume (m³)	Group	Shell thickness (cm)	Shell mass (kg)
1A	Baseline	132.1	92.7	All	0.919	3,300
1B	Baseline, varied element thickness	132.1	92.7	Tunnel Hatch area Cylinder End dome	0.841 0.978 0.267 0.612	1,954
2A	Expanded baseline	161.9	92.7	All	0.848	3,429
2B	Expanded baseline, varied element thickness	161.9	92.7	Tunnel Hatch area Cylinder End dome	0.765 0.932 0.254 0.584	1,973
3A	Sphere	142.5	79.8	All	0.574	2,059
3B	Sphere, varied element thickness	142.5	79.8	Tunnel Hatch area Sphere	0.650 0.495 0.152	803
4A	Torus	90.13	81.1	All	0.432	1346
4B	Torus, varied element thickness	90.13	81.1	Tunnel Hatch area Torus	0.493 0.419 0.152	637
5A	Toroid	159.5	91.2	All	0.866	3,683
5B	Toroid, varied element thickness	159.5	91.2	Tunnel Hatch area Outer wall Inner wall End dome	0.782 0.965 0.198 0.152 0.152	995
6A	0.707	158.4	91.2	All	0.605	2,266
6B	0.707, varied element thickness	158.4	91.2	Tunnel Hatch area Cylinder End dome	0.546 0.630 0.152 0.152	755
7A	Hotdog	124.8	79.5	All	0.566	1,857
7B	Hotdog, varied element thickness	125	79.5	Tunnel Hatch area Cylinder End dome	0.483 0.577 0.229 0.152	1,857

[a] Adapted from Bodkin *et al.* [4].

thereby permitting areas of low stress to have a thinner wall. Following analysis of the A and B cases it was determined that optimizing material thickness throughout a geometry afforded the most weight savings. Deciding upon the best module was determined by trading design aspects such as volume and weight. In deciding on a module the team was looking for the highest livable volume and lowest mass-per-volume ratio. For example, Case 1B clearly has the highest livable volume but also the highest mass-per-volume ratio of the varied thickness cases, whereas the Hotdog has the lowest mass-per-volume ratio of all the varied element thickness cases but suffered in the analysis because it offered the lowest livable volume. Ultimately, the trade studies recommended Case 6B as the desired module, a design that offered $158\,\mathrm{m}^3$ total volume and $92.1\,\mathrm{m}^3$ livable volumes [4].

Habitat functionality

The next task the OSC team tackled was characterizing the *functionality* of the Habitat Module. Functionality is a criterion that defines the module's design constraints and also describes the tasks it can support at any stage in the outpost's development schedule. One suggestion for the initial outpost configuration was a base consisting of four habitation modules, as described in Table 5.3.

Radiation shielding

Having defined the Habitat Module's geometry and functionality, OSC's attention turned to the issue of radiation protection. Shielding a lunar base from the hazards of radiation is a critical technology since the safety of each crewmember depends on it. Many people think the Apollo astronauts, who didn't have much protection, were fine but the two-to-three-day Apollo missions were very short compared with the future six-month duration increments planned by NASA.

Materials capable of effectively shielding astronauts against space radiation are based on their nuclear, atomic, and molecular cross-sections, of which the latter two are dependent on the density of electrons per unit volume and the electron density. This means the most effective radiation shielding material is the one with the highest electron density. Since hydrogen has the highest electron density of any atom it also possesses the most effective shielding characteristics. Another advantage of hydrogen is its ability to cause fragmentation of radiation particles without causing secondary particles through fragmentation of the shielding material nuclei, as it cannot fragment into another nucleus. However, although a material possessing high hydrogen content will normally have superior shielding properties, it will not necessarily possess the high structural integrity required for habitat construction. For example, lithium hydride, a material with high hydrogen content and one often used as a shield for nuclear reactors, possesses the shielding effectiveness required to protect astronauts on the lunar surface. Unfortunately, the material does not lend itself to the requirements of building a habitat as it cannot perform the multiple functions required by habitat engineers.

Presently, NASA uses aluminum as radiation shielding, a material that is not very protective due to its low electron density. Aluminum's days may be numbered,

Table 5.3. Habitat functionality [4].

Habitat Module	Module functionality	Module systems
Habitat Crew Module	Supports four crewmembers Crew quarters Galley and dining Hygiene Laundry Health	ECLSS management Power management Thermal control Communications
Habitat Science Module	Science Exercise	Power management Thermal control Communications
Habitat Maintenance Module	Dust control Maintenance Storage Hygiene	Power management Thermal control Communications
Habitat Logistics Module	Storage Maintenance	Power management Thermal control Communications

however, because recently polymers have attracted the interest of engineers. Due to the high hydrogen content of polymers and because so many variants of the material exist, it would seem this multi-functional material may be the radiation shielding engineers have been searching for.

Polyethylene (PE) has been singled out because it inherently has the highest hydrogen content of any polymer. Also, PE does not contain any large nuclei, important because the absence of large nuclei dramatically reduces the risk of the shielding material fragmenting as a result of collisions with radiation ions.

Studies comparing the effectiveness of PE against the shielding properties of other materials have demonstrated that PE, and PE combined with epoxy, are very similar in their shielding effectiveness.

Polyethylene and RFX1

PE combined with epoxy provides excellent radiation shielding and is stronger than PE alone, but the material is still not sufficiently rigid to conform to the very demanding standards of Lunar Habitat designers. Recently, however, Nasser Barghouty, Project Scientist for NASA's Space Radiation Shielding Project at the Marshall Space Flight Center (MSFC), and Raj Kaul (Figure 5.1), a NASA researcher, have developed a PE-based material called *RFX1*. RFX1 is both lighter and stronger than aluminum, thereby providing habitat designers with a material combining superior and shielding properties. In fact, compared with aluminum, RFX1 is 50% more effective at shielding against SPEs and 15% more effective against

Figure 5.1. NASA researcher, Raj Kaul, examines a radiation "brick" of RFX1, a material that may offer radiation protection for astronauts on the Moon. Image courtesy: NASA.

GCRs. Although it is a novel material in the space industry, RFX1 has already been used in the military as protective armor for helicopters, making it an obvious choice as a ballistic shield against micrometeorites.

Since RFX1 is a fabric, the material can be draped around molds and shaped into specific spacecraft components. Another advantage of RFX1 is that it produces less secondary radiation or secondary emissions than heavier materials such as aluminum. Secondary radiation occurs when particles of space radiation collide with atoms in the shield, resulting in small nuclear reactions that may damage cellular structure. Although most people assume lead provides the best radiation protection, this is not the case because, although lead does soak up a lot of radiation, it also produces significant amounts of damaging secondary radiation.

The task of designing a shield suitable for lunar missions is a complex process, even if engineers are provided with accurate information pertaining to space radiation sources and biological response functions to radiation exposure. Current knowledge regarding these parameters is not presently available so, in the absence of such data, NASA engineers referred to a worst-case scenario based on the February 1956 SPE, the most dangerous radiation event on record. Engineers then analyzed how much PE would have been required to mitigate such an event. After crunching the numbers it was determined 19.05 cm of PE would have been required to protect the crew, but since such a thickness of PE would result in an addition of 6,804 kg of weight to Habitat Modules, the concept of merely adding PE was immediately discounted. Instead, engineers conducted more research evaluating the use of *in situ* resources to provide radiation shielding and determined 50.8 cm of regolith would have mitigated the February 1956 SPE. However, erring on the side of caution engineers designed a habitat that could protect crewmembers from a SPE four times greater than the February 1956 SPE, eventually resulting in a design requiring 203 cm of regolith coverage.

Launch manifest

The OSC team had now addressed the issues of module functionality, geometry, and radiation shielding. The next matter was deciding how the modules would be dispatched to the lunar surface. Although the first lunar missions are almost a decade away, NASA has published a lunar launch manifest starting in 2017 with the launch of surface infrastructure assets. The Habitat Maintenance Module (HMM), the first of the infrastructure assets, will be followed by the Habitat Crew Module (HCM), the Habitat Logistics Module (HLM), and finally the Habitat Science Module (HSM). Since the HMM provides sufficient habitable volume for an extended stay (~14 days), it will serve as a base for the first crew of four and, as other modules are added, the habitable volume and functionality of the lunar base will slowly increase, thereby permitting longer duration stays [4].

Module layout

Using NASA's lunar launch manifest as a guide, the OSC team considered the best way to join the habitats, a task resolved by yet another trade study. First, the team agreed upon a set of ground rules. It was agreed each configuration should include four airlocks so, in the event of an airlock failure, crewmembers would be able to access another module. The OSC team also decided each configuration should maximize the number of cargo container (CC) interfaces.

After agreeing upon the ground rules, five module configurations were evaluated through trade studies assessing the significant characteristics of each module (Table 5.4). The straight line design, for example, was immediately discounted due to safety concerns in the event of an obstruction of the module's only exit/entrance [4].

Following the trade studies, the most versatile, safe, and logical design was the *staggered* configuration since it made the most effective use of the available area and permitted a high number of airlocks.

Habitat surface power

The next step was to define primary and back-up power systems for lunar base operations. Since a power failure could result in loss of mission and jeopardize the crew, it was essential a lunar surface power system be able to provide power in both nominal and emergency situations.

The hostile nature of the space environment is not kind to engineering systems and subsystems, hence the requirement for redundancy. Before considering the type of back-up power system, engineers needed to decide how much power would be required, what type of power to use, and where to locate the power plant.

Initially, mounting solar cells on the habitat walls was considered. This concept was rejected due to the problem of regolith accumulating on the solar cells during the

Table 5.4. Lunar Base configuration assessment [4].

Configuration	Characteristics	Assessment
Square	• Simple layout with four airlocks • Permits up to eight CC habitat dockings • Permits up to three CC airlock dockings • Failure of single airlock limits access from a Habitat Module to only one other Habitat Module via available airlock • Failure of single Habitat Module limits access to other modules	Inability to provide crew with multiple egress routes during an airlock, or habitat failure eliminates this configuration from consideration
In-line	• Simplest layout; four airlocks; design assumes single airlock door used for habitat ingress/egress • Permits up to nine CC habitat dockings • Permits up to eight CC airlock dockings • Failure of single airlock/Habitat Module prevents access to other Habitat Modules	Inability to provide crew with multiple escape routes during an airlock or Habitat Module failure eliminates this configuration from consideration
Right triangle	• Complex layout; five airlocks; design assumes single airlock door used for habitat ingress/egress • Permits up to six CC habitat dockings • Permits up to five CC airlock dockings • In most cases, failure in single airlock permits access from a Habitat Module to more than one Habitat Module via other available airlock	In most cases, access to other Habitat Modules is not restricted if an airlock or Habitat Module failure occurs
Star	• Complex layout; four airlocks; design assumes single airlock door used for habitat ingress/egress • Permits up to five CC habitat dockings • Permits up to five CC airlock dockings • In most cases, failure in single airlock permits access to more than one Habitat Module via other airlock • In all cases, failure of single Habitat Module does not limit access to other Habitat Modules to only one airlock	Capability to provide unrestricted access to other Habitat Modules in the event of a Habitat Module failure is beneficial
Staggered	• Complex layout; four airlocks; design assumes single airlock door used for habitat ingress/egress • Permits up to six CC habitat dockings • Permits up to five CC airlock dockings • In most cases, failure in single airlock permits access to more than one Habitat Module via the other airlock • In all cases, failure of single Habitat Module does not restrict access to other Habitat Modules to only one airlock	Preferred configuration

placement of regolith as radiation shielding. Instead, OSC decided the primary power source should be a nuclear power plant, buried one kilometer away from any crewed activity to reduce radiation exposure [3, 4]. The reactor back-up, intended to serve as the initial power source for the first year of operations, will consist of batteries. In case of battery failure, habitat power will be supplied by a regenerative fuel cell (RFC)–solar array combination. In the event both systems fail, a tertiary level of redundancy ensures each habitat has sufficient emergency power to allow crew-members to don EVA suits and evacuate the habitat. In both the nuclear and battery system, the power will be supplied to surface assets via a series of 5 km long cables and power nodes, the latter providing access to surface mobility assets for recharging.

LUNAR LIFE SUPPORT

One of the greatest challenges posed by extended-duration missions on the lunar surface is ensuring basic life support requirements can be met for long periods of time. The Moon's remote location, combined with its challenging surface environ-ment and reduced gravity level, presents unique challenges in terms of designing life support methods. Compounding the challenge is the limited experience in the field of extended-duration life support systems and subsystems which, to date, is restricted to submarines, the ISS, Space Shuttle and Earth-based biospheres.

The lunar Environmental Control and Life Support System (ECLSS) will be selected by an evaluation of mission parameters such as mission duration, crew size and composition, crew tasks, daily activities schedule, and overall mission timelines. Having considered these factors, trade studies have been and continue to be con-ducted to determine the level of reusability in the system, a function described by the terms *open* loop and *closed* loop.

An *open-loop system* is one containing no reusable features. Such a system must be constantly restocked, whereas in a *closed-loop system* each resource is recycled. Due to the different mission profiles of the Constellation Program it was not possible to choose one system for all missions. For example, short-duration missions favor the open-loop system since it has the benefit of being simple, highly reliable, and more mass-efficient. The closed-loop system, with its increased system mass, is more appropriate for long-duration missions due to the extra subsystems required for recycling resources. To determine the advantages and disadvantages of each ECLSS, another trade study was conducted which revealed a closed-loop system saved three missions every year thanks to mass savings. However, since the first mission is 14 days long, the most appropriate ECLSS is the open-loop system since it is more mass-optimized. Therefore, for the initial 14-day mission, a CC will be used to transport sufficient oxygen and nitrogen for 60 crew-days. Gradually, as missions become longer, elements of a closed-loop system will be added to the base, although, since the efficiency of these systems has yet to reach 100%, they must be considered a partial closed-loop system.

Regardless of which system is used, it must be capable of providing the crew with a comfortable atmosphere that complies with the parameters described in Table 5.5.

Figure 1.3. Orion is the name of NASA's Crew Exploration Vehicle. Image courtesy: NASA.

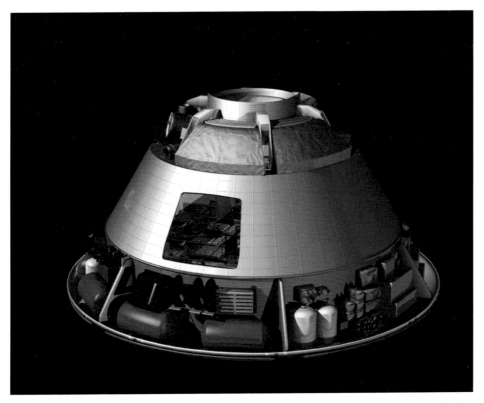

Figure 4.6. Orion capsule showing RCS system and propellant tanks. Image courtesy: NASA.

Figure 4.12. Orion's Launch Abort System. Image courtesy: NASA.

Figure 5.6. NASA's Mobile Lunar Transport, seen here undergoing testing at Moses Lake, Washington, in June 2008, is a highly mobile concept lunar truck. Each set of two wheels can pivot individually in any direction, giving the rover the ability to drive sideways, forward, backward, and any direction in between. Image courtesy: NASA.

Figure 5.7. The All-Terrain-Hex-Legged Extra-Terrestrial Explorer (ATHLETE) is capable of rolling over undulating terrain. The image shows the vehicle carrying a pressurized habitat during testing at Moses Lake, Washington, June 2008. Image courtesy: NASA.

Figure 6.2. Ernest Shackleton. Image courtesy: NOAA.

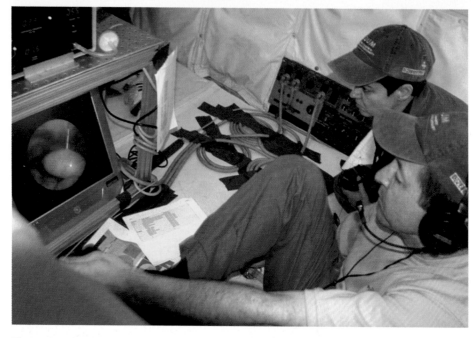

Figure 6.6. Dr. Scott Dulchavsky views a laparoscopic image of a surgical procedure being performed remotely by a non-physician on the NASA Microgravity Research Facility. He uses just-in-time training, combined with voice commands, to guide a minimally trained operator to perform complex procedures in space. Image courtesy: NASA.

Figure 6.7. Telemedicine may require astronauts to use smart medical systems similar to the one depicted in this photo. The training session shown here highlights the basics of an ultrasound examination such as probe positioning, the location of the organ within the body, the size and structure of the organ, and what the correct ultrasound image should look like on the monitor. Courtesy Dr. Scott Dulchavsky, NSBRI.

Figure 9.3. Demonstrating the Mark III suit's dexterity, an astronaut captures a scene at Moses Lake, Washington, during a robot demonstration in June 2008. Image courtesy: NASA.

Figure 9.4. Spacesuit engineer Dustin Gohmert simulates work in a mock crater of JSC's Lunar Yard, while his mode of transport, NASA's prototype lunar rover stands in the background. The rover has the ability to lower itself all the way to ground, making climbing on and off easy, even in the Mark III spacesuit. Image courtesy: NASA.

Figure 9.5. The revolutionary Bio-Suit. Image courtesy: Professor Dava Newman, MIT: Inventor, Science and Engineering; Guillermo Trotti, A.I.A., Trotti & Associates, Inc. (Cambridge, MA): Design; Dainese (Vicenza, Italy): Fabrication; Douglas Sonders: Photography.

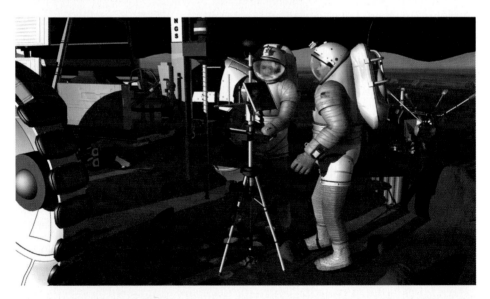

Figure 9.6. Artist's rendering of two astronauts establishing a worksite with scientific equipment. The astronauts install the equipment and then radio to the outpost to confirm all systems are operational. The crewmembers then activate the systems and data are sent to Earth. Image courtesy: NASA.

Table 5.5. Typical Space Habitat atmosphere requirements [16].

Atmosphere parameter	Nominal value
Total pressure	99.9–102.7 kPa
Partial pressure of oxygen	19.5–23.1 kPa
Partial pressure of nitrogen	79 kPa
Partial pressure of carbon dioxide	0.4 kPa
Temperature	18.3–23.9°C
Relative humidity	30–70%
Ventilation	0.076–0.203 m/s

Human life support requirements

The primary *consumables* (often referred to as *inputs*) required by the crew are oxygen, water, and food, whereas the primary *outputs* are carbon dioxide, urine, and feces. To determine input and output amounts, mission planners calculate supply cycles because regardless of the type of Life Support System (LSS) or environment, the consumable supply cycles of atmosphere, water, and food ultimately determine the design of a LSS. Given the impact of these supply cycles, it is useful to examine the daily inputs and outputs of an astronaut, based on previous spaceflight experience, in order to generate a representative estimate of the human life support consumable mass requirements (Table 5.6).

Habitat atmosphere considerations

The data presented in Table 5.6 provides mission planners and engineers with baseline input and output requirements, information used to ensure the atmosphere of the Lunar Habitat meets the demands of crew comfort.

One of the most important considerations that must be incorporated into the ECLSS design is a means of carbon dioxide removal since each crewmember produces 1.0 kg of carbon dioxide daily. However, in addition to removing carbon dioxide, engineers must also ensure some means of generating oxygen. Other parameters engineers must carefully consider include the operating pressure and air mixture ratio. From an engineering perspective it is desirable to maintain the habitat at a low operating pressure, not only to limit the time required for the pre-breathe prior to surface EVA, but also to limit leakage rates in the event of an emergency. However, the lower the total operating pressure, the greater the concentration of oxygen required to maintain the partial pressure of oxygen (PPO_2). This causes a potential problem since, if the PPO_2 is too low, the crew risk suffering from hypoxia. Hypoxia

Table 5.6. Daily human input and output requirements.[a]

Life Support System inputs		Life Support System outputs	
Input	Average amount (kg)	Output	Average amount (kg)
Oxygen	0.84	Carbon Dioxide	1.00
Food solids	0.62	Respiration and perspiration water	2.28
Water in food	1.15	Food preparation, latent water	0.036
Food preparation water	0.76	Urine	1.4
Drink	1.62	Feces water	0.091
Metabolized water	0.35	Feces solid	0.018
Hand/face wash water	4.09	Sweat	1.4
Shower water	2.73	Hygiene water	12.58
Urinal flush	0.49	Clothes wash water Liquid Latent	11.90 0.60

[a] Values based on mean metabolic rate of 136.7 W/person and respiration quotient of 0.87 [12].

is a deleterious condition brought about by insufficient oxygen to support physiological functioning and a crewmember suffering from the condition may exhibit symptoms ranging from hallucinations and paralysis to loss of memory and death. It would seem the solution would be to add more oxygen to the system but this creates another safety concern by increasing atmosphere flammability. Obviously, the aim of engineers is to reduce the flame spread rate in the event of fire, but this can only be achieved by reducing the oxygen content!

In the OSC study, the oxygen issue was resolved by performing more trade studies, leading to the compromise of an air composition of 35/65 O_2/N_2 ratio, resulting in an acceptable flame spread rate.

Habitat temperature and humidity

Temperature regulation within the habitat presents yet another significant challenge to engineers as the outpost will be subject to the extraordinary temperature extremes described earlier. The challenge of ensuring a stable temperature within the habitat is compounded by other factors implicated in the thermal regulation of an enclosed volume. Engineers must not only consider outside temperature differences but must also take into account the heat generated by equipment and human metabolism in addition to ensuring humidity levels remain tolerable for humans and plants without leading to electrical shorts!

Energy expenditure

Once the atmosphere considerations have been addressed the next issue is the energy expenditure of the crewmembers since this factor will influence the amount of consumables that must be either brought from Earth or produced *in situ*. Although the baseline input and output values described in Table 5.6 are useful, engineers also need to know how much energy astronauts are likely to use during normal daily lunar activities such as surface excursions. As an example, an estimate of the energy expended by a crewmember performing a demanding EVA day is provided in Table 5.7.

Similar calculations have been estimated for other anticipated daily lunar workloads such as a lunar rest day and a ten-hour intravehicular activity (IVA) work day which might involve performing necessary laboratory work. Based on these calculations, nutritionists are able to calculate how much food will be required per astronaut per day. While on the lunar surface, the energy supply of astronauts will be comprised of 50%–55% carbohydrates, 20% lipids, and between 20% and 25% proteins [6].

Table 5.7. Energy expenditure during 10-hour lunar EVA [7].

Daily activity	Daily time spent (h)	Energy expenditure (kcal/kg/min)	Daily activity, energy expended (kcal)
Off-duty			
Sleeping	8	0.015	511
Meal prep/eat	2	0.04	341
Personal Hygiene	1	0.03	128
Reading/Sitting	1	0.022	94
Housekeeping tasks	1	0.06	256
Exercise	1	0.115	490
Total off-duty	14		1819
On-duty			
Walking	3	0.07	895
Standing	2	0.03	256
Kneeling	2	0.025	213
Crouching	2	0.04	341
Digging	1	0.12	511
Total on-duty	10		2,215
Total on-duty + off-duty	24		4,034

If mission planners wanted to be cruel, they could design a pre-packaged food inventory similar to the food provided to long-stay crewmembers onboard the ISS, amounting to a mass of 1.83 kg per crewmember per day. However, food variety will be necessary to sustain crew morale, especially when the crew will be living in a habitat nearly 400,000 kilometers from home. Most likely, NASA nutritionists will be generous and plan for a sustainable diet that relies on carbohydrates and salad from food crops, a plan that will result in a mass of between 3.6 kg and 3.82 kg per crewmember per day [14].

ECLSS technology options

The previous sections addressed the human factors that engineers must address when designing the habitat's ECLSS. To ensure the habitat atmosphere is maintained and regulated, potable water management and waste water management are provided, and sufficient food is available to crewmembers, engineers have two technology options available. The first of these is the *physico-chemical*/closed loop, which uses fans, filters, and chemical separation processes, and the other is *bioregenerative*, which utilizes living organisms such as plants [19]. Since physico-chemical technology is well understood, compact, and easy to maintain, space missions have historically used this method of closing the loop. However, this method has only been possible because all missions in the last three decades have been in close vicinity of Earth so the penalties of storage and transportation and the problems of resupply have not been critical issues. However, astronauts inhabiting a permanent settlement on the Moon will be unable to replenish stocks without significant transportation cost. To combat these problems it is likely the macro Life Support System of the Lunar Habitat will comprise a hybrid physico-chemical/bioregenerative design [13], which is described in the following section.

Design of a physico-chemical/bioregenerative ECLSS

Atmosphere regulation

An Atmosphere Regulation and Monitoring System consists of subsystems responsible for controlling pressure, temperature, humidity, and ventilation. Pressure control is normally achieved by using control algorithms, using data provided by pressure sensors that are in turn connected to valves, regulators, and heaters. Temperature control is normally achieved by using a water coolant loop, used to transfer internal and external heat. Humidity removal is ensured by reducing the air temperature below its dew point and separating the resultant condensed water from the air flow, and ventilation of the habitat is achieved by the use of fans, isolation valves, and ducting.

To enable some of these functions in the closed-loop system envisaged for the Lunar Habitat an Air Revitalization System (ARS) will be required. The ARS consists of several independent air loops that circulate habitat pressure atmosphere. Such a system ensures humidity remains within prescribed levels, ensures carbon dioxide and carbon monoxide levels remain non-toxic, temperature and ventilation are regulated, and the habitat's electronics are cooled. Other ARS subsystem functions include removing carbon dioxide and generating oxygen. Each of these functions may be achieved through either a physico-chemical or a bioregenerative process.

Working in conjunction with the ARS will be an Atmosphere Revitalization Pressure Control System (ARPCS). The ARPCS will ensure habitat air pressure is maintained within prescribed levels and partial pressures of oxygen and nitrogen are also maintained within nominal levels. In the ARPCS, oxygen and nitrogen supply systems are controlled by an atmosphere pressure control system that regulates gas release through an arrangement of check valves, inlet valves, relief valves, supply valves, sensors, control switches, and talkback systems.

The ARPCS also performs a critical function in the event of a rapid decompression. In the event habitat pressure falls below or rises above prescribed levels, or if PPO_2 falls below or rises above nominal levels, a master alarm will sound and red caution lights will illuminate on the ARPCS panel. Whichever crewmember is closest to the ARPCS panel will be responsible for responding to the emergency, which will require a thorough knowledge of the location of the habitat's overpressure and negative relief valves. For example, if the emergency is an over-pressurization situation, the crewmember will need to activate the habitat relief switch which will in turn activate a motor-operated valve designed to relieve pressure through venting. Conversely, if the emergency is a low-pressure alarm, the crewmember will need to activate the negative pressure relief valves that will cause a flow of ambient pressure into the habitat.

In addition to the ARS and ARPCS the habitat will be fitted with an Active Thermal Control System (ATCS), the ECLSS component responsible for heat rejection. The ATCS performs its function through an arrangement of cold-plate networks, coolant loops, liquid heat exchangers, and various other heat sink systems that reject heat outside the habitat. Since the habitat will feature a large number of electronic units and systems that generate heat, crewmembers will have to be careful not to overload the heat sink systems. If excessive heat does build up and the capacity of the heat sink units is exceeded, the habitat will have the capability to activate a flash evaporator designed to meet excess heat rejection requirements for short periods.

Physico-chemical processes

A typical physico-chemical process for removing carbon dioxide is to use lithium hydroxide (LiOH). LiOH is more commonly used in breathing gas as a purification system onboard submarines and closed-circuit re-breathers to remove carbon dioxide from exhaled gas by producing lithium bicarbonate [10]. In an enclosed habitat, LiOH may also be used as a means of reducing carbon dioxide, while at the same time regenerating oxygen. This process is achieved physico-chemically by a chemical reaction such as the Sabatier process which uses a catalytic methanation reaction between carbon dioxide and hydrogen at high temperatures to generate water vapor.

The water vapor can be recovered by passing product gases through a condensing heat exchanger and the water generated can be electrolyzed to produce oxygen [11].

Bioregenerative processes

Bioregenerative processes are often referred to as Biological Life Support Systems (BLSSs). BLSSs work by integrating carbon dioxide assimilation, carbon dioxide reduction, and oxygen generation together with other bioregenerative processes for food production and treatment of waste. Major components of bioregenerative processes include photosynthetic organisms such as plants and algae which produce food and oxygen and at the same time remove carbon dioxide from the habitat atmosphere. However, bioregenerative processes do not represent a stand-alone system since physico-chemical subsystems are required to support certain biological functions such as temperature and humidity. Also, in order to convert biomass into edible food and to convert waste products such as waste water into useful resources, other physico-chemical subsystems must be used [8].

While growing plants may seem to represent an elegant bioregenerative process, the input requirements in terms of light energy, carbon dioxide, and water may be quite significant. Algae, on the other hand, also have the ability to perform photosynthesis, assimilate carbon dioxide, and release oxygen. However, unlike most plants, algae, by virtue of their unique unicellular structure, are able to convert light energy into biomass with greater efficiency than plants and therefore confer significant advantages with respect to atmosphere regeneration and harvest index. Unsurprisingly, given all the advantages of algae, several studies have been conducted to determine the most suitable candidates for a lunar BLSS. After many studies the two most promising algae candidates under consideration include green algae (Chlorophyta) and blue-green algae (Cyanophyta) such as *Cyanobacteria*, *Spirulina*, and *Anacystis*. Recently, particular research attention has been paid to *Spirulina*, due to its high photosynthetic qualities and its ability to use very few byproducts in its own growth, which means that a potential may exist for using *Spirulina* as a nutritious food source [2].

Water on the Moon

A limited number of options exist for ensuring the crew has sufficient water. The first option is re-supply but obviously this is not only extraordinarily expensive, it also represents a major logistical headache.

The next option is the physico-chemical one in which water evaporated through normal environmental processes is recovered using a Water Condensation System. The recovered water, which could also include urine, would then be used for oxygen generation or simply recycled as potable water. Extracting water from urine would require a little more work since it would first need to be centrifuged, evaporated at low pressure, condensed, and finally subjected to a high-temperature catalytic reactor assembly which would remove any organic contaminants.

Another option is to use the Sabatier process mentioned earlier, or to use a filtration system in which waste water is multi-filtered through special membranes before being partially oxidized and ion-exchanged to make it potable.

Yet another system that may be used is the Vapor Phase Catalytic Ammonia Removal (VPCAR) System which distills wastewater using wiped film rotating disks (WFRD) that remove contaminants. In this system, oxidation and reduction reactors treat the distillate to oxidize organic elements and to reduce oxidized nitrogen compounds to nitrogen gas. A highly integrated unit, the VPCAR system has achieved 98% water recovery and therefore is a likely candidate for inclusion in a Lunar Habitat ECLSS.

The bioregenerative water option achieves water regeneration by using biological reactors to oxidize organic compounds. The process starts with waste water flowing into plant fields where it is absorbed by algae and transpired. Air conditioners, phytotron condensers, and a drying chamber condense the transpired water and an incinerator is then used to burn inedible biomass in order to extract purified water. Finally, remaining contaminants are destroyed using ion exchange beds and photo-oxidation. Like the VPCAR system, the bioregenerative option achieves 98% water recovery.

Food

Re-supply of food for a long-duration mission will probably not be considered as an option for two reasons. First, the extra launches required to supply food increases the cost of the Lunar Habitat significantly, and, second, the habitat would need to set aside areas for food storage and packaging the resultant waste. Food production methods therefore, such as algal, plant, and animal systems, will probably be used to sustain crewmembers during six-month missions.

One of the most efficient food production systems is algal since algae grow rapidly, produce oxygen, and certain types can be eaten, although this latter benefit may not be deemed an advantage by the crew! However, if the "algae as a food source" idea is rejected by the crew it will still be possible to utilize the algae as a fertilizer or as a food source for plants. The use of algae as a food source for plants will be important as plants will most likely provide a majority of the crew's nutritional requirements. Careful selection of plants will ensure the crew is supplied with most of their daily requirement of proteins, carbohydrates, fats, minerals, and vitamins, and as a side-benefit, plants will also contribute to oxygen generation and may be utilized for water purification.

It has been suggested the crew might supplement their diet with animal food sources, but initially this will not be possible due to the added energy, mass, and volume required by the habitat system, a requirement that also increases the habitat's complexity. Once the base has been established, however, it is likely animals, particularly those that confer stability to the system and those with low mass and short gestation periods, may provide an additional food source for the crew.

Waste

On the ISS, waste is simply stored in Progress capsules until the capsules are full, at which point they are de-orbited and re-enter the Earth's atmosphere. Unfortunately, no such convenient solution will be available to astronauts living on the lunar surface but there are other options for treating the different types of crew waste.

Gray-water treatment

Washing dishes, taking a shower, and using the urinal each contribute to gray-water waste which can be treated by passing it through an ion exchange bed (IEB), thereby eliminating the mineral content. Once the water has been de-mineralized it is processed and purified by micro-algae inside an algal cultivator compartment, which also recycles the atmosphere by processing the carbon dioxide expelled by the crew and producing oxygen.

Black-water treatment

Water contained in food wastes and human feces is considered *black water* and requires a more aggressive treatment process than that used to treat gray water. The treatment involves several stages, first utilizing proteolytic bacteria to anaerobically transform the black waste into hydrogen, carbon dioxide, fatty acids, minerals, and gray water. Outputs from the black water are then processed by a photoheterotrophic device (one that uses bacteria that manufacture food using light energy) that uses *Rhodospirillum rubrum* bacteria to eliminate terminal products, ultimately generating more water, although at this stage the water still needs to undergo further processing. The next stage converts ammonium derived from biological waste into nitrate, which can then be used as a nitrogen source for plants. Finally, the water generated by the photoheterotrophic device is sent to an algal cultivator where it follows the same cycle as gray water.

Contamination design considerations

An important design aspect of the habitat will be to provide energy-efficient methods of decontamination in order to maintain base hygiene and safety. For example, astronauts and cosmonauts onboard Mir discovered certain mutant strains of bacteria could attack rubber seals and Apollo astronauts were plagued by the lunar dust that covered not only their EVA suits but practically every surface inside the LEM. Since astronauts will be returning samples to the habitat, methods of encapsulation, containerization, and sterilization will need to be developed.

At the time of writing, the ECLSS described has proven effective in terms of closing the air, water, and waste loops, but the nutritional requirements are not completely met, which means that the crew will require nutritional supplements that may be provided by crew rotation or logistics missions.

Fire detection and suppression

Smoke and fire detection and suppression capabilities will be provided throughout the Lunar Habitat by means of ionization detection elements that provide information concerning smoke concentration levels to the performance-monitoring general purpose computer (GPC). In the event of the GPC detecting an abnormal concentration of smoke, the master alarm's red lights will activate and the general alarm will sound throughout the habitat. Once the fire has been detected, crewmembers will attempt to isolate the fire and use fitted systems such as halon-1301 to fight the fire. In the event of the fire being uncontained, the crew will need to evacuate to a safe haven and depressurize the habitat in order to extinguish the fire. Once the fire has been extinguished, crewmembers will then re-enter the habitat and check for toxic products, before repressurizing. The subject of crew emergency training, which includes responses to fires, is described in Chapter 6.

EXPANDABLE HABITAT STRUCTURES

Using information generated by the OSC studies and other companies, NASA has begun to develop concepts for habitation modules. Perhaps the most popular option presently being studied is the use of inflatable structures.

In addition to being easily compressed and cheap to transport, inflatable structures provide large living spaces and have already been demonstrated in the harsh environment of low Earth orbit (LEO). In July 2006, Bigelow Aerospace launched Genesis I which, until the launch of Genesis II (Figure 5.2) the following year, represented the most successful use of an inflatable structure in space.

The inflatable design was initially developed by NASA's TransHab Project which first proposed the inflatable module concept as a crew quarters for the ISS. Inside the Space Shuttle's cargo bay the module would have had a diameter of 4.3 meters, but once inflated it would have expanded to 8.2 meters, providing the crew with a 340 cubic meter facility. The TransHab Module (Figure 5.3) originally designed for the ISS was to have been a unique hybrid structure that combined the mass efficiencies of an inflatable structure with the advantages of a load-bearing hard structure. The inflatable shell consists of multiple layers of blanket insulation and successive layers of Nextel designed to break up particles of space debris and micrometeoroids that may impact the shell at speeds seven times faster than that of a bullet.

Unfortunately in 2001, due to the $4.8 billion ISS budget deficit, Congress restricted NASA from spending any money on TransHab beyond the design studies. After NASA abandoned the TransHab technology, Robert T. Bigelow's company, Bigelow Aerospace, decided to refine the idea. Bigelow started by purchasing a building near Ellington Field in late 2004 and, with the help of employees of Johnson Space Center's (JSC's) Structural Engineering Division, transformed the building into an inflatable production facility. Bigelow also licensed the rights to the patents for several TransHab technologies and materials and arranged Interpersonal Act Agreements with JSC's inflatable module experts, Jason Raboin, Chris Johnson,

Figure 5.2. Bigelow Aerospace's Genesis II inflatable habitat passes over the Baja Peninsula. Image courtesy: Bigelow Aerospace.

Gary Spexarth, and Glenn Miller, permitting them to work full-time with Bigelow Aerospace. The culmination of Bigelow's work was the flights of Genesis I and II, demonstrating the potential of multipurpose habitable space structures. After the success of Genesis I and II, Bigelow envisions a privately owned orbital facility where the inflatable space structures could be assembled before being flown directly to the lunar surface. NASA, having seen the success of Bigelow's efforts, is encouraging the company in its development of the inflatable structures with a view to possibly using them as habitats on the Moon.

Until Bigelow's inflatable structures are developed for lunar applications, NASA is pursuing other inflatable options. Recently, NASA, the National Science Foundation, and ILC Dover of Frederica, Delaware, joined forces to investigate the potential of inflatable structures as long-term habitats for lunar astronauts. In January 2008, an inflatable Lunar Habitat structure was erected at McMurdo Station, Antarctica, for the purposes of testing the integrity of the habitat in an extreme environment,

Figure 5.3. A TransHab module undergoing testing. Image courtesy: NASA.

analogous to the lunar surface. The habitat (Figure 5.4), which weighs less than 500 kg, consists of a tubular inflatable structure, an insulation blanket, power and lighting systems, heaters, a pressurization system, and a protective floor. While the habitat is deployed in Antarctica, engineers will have the opportunity to research ways to apply regolith to the walls of the structure for radiation shielding and investigate the efficacy of dust mitigation strategies.

Although the word "inflatable" doesn't conjure up images of cutting-edge spaceflight technology, given all the advantages of inflatable structures it is likely this habitat design will be the module of choice when astronauts finally return to the Moon. When that time arrives, it is almost certain private companies such as Bigelow Aerospace will play a vital role in the habitat's construction and deployment.

Figure 5.4. NASA, the National Science Foundation and ILC Dover unveiled an Antarctic-bound inflatable habitat on November 14, 2007. Image courtesy: NASA.

LUNAR COMMUNICATIONS AND NAVIGATION CAPABILITIES

Communication and navigation overview

Outpost communications and navigation (C&N) services will be provided via relays and surface lunar communication terminals (LCTs) with periodic direct to Earth (DTE) capability. One component of the system will be a lunar relay satellite (LRS) that will provide coverage of the outpost and the rest of the Moon with communication, tracking, and time services via the LCT.

Concept of operations

During surface missions, outpost information will be routed via the LCT to other lunar users or to Earth via DTE or LRS. The combined capabilities of the LRS and LCT will provide services such as forward and return voice, video, one-way and two-way ranging, in addition to providing fully routed data between Earth, lunar orbit, and lunar surface users, as described in Table 5.8.

Descent and landing navigation capability

Installed near the outpost will be an autonomous landing and hazard avoidance technology (ALHAT)-based infrastructure comprised of a passive optical system

Table 5.8. Outpost communication traffic model.[a]

Description	System	Data rates		
		Low rate (Mbps)	High rate (mbps)	Total rate (mbps)
Aggregate peak rate to Earth	LRS and Earth Ground System	3.9	151.0	154.9
Aggregate peak rate from Earth	LRS and Earth Ground System	1.1	66.0	67.1
Aggregate peak rate up to LRS from lunar surface	LRS and LCT	6.4	216.0	222.4
Aggregate peak rate down from LRS to lunar surface	LRS and LCT	6.1	141.0	147.1
Aggregate peak rate across lunar surface	LCT	8.7	143.0	151.7

[a] Adapted from NASA's *Lunar Communications and Navigation Architecture Technology Exchange Conference, November 15, 2007.*

and strobe lights designed to be used in the last 300 m of descent in low-light landing situations. The ALHAT-based infrastructure will provide crewmembers landing on the Moon with the capability to maintain an accurate trajectory throughout the powered descent regime and the ability to land within one meter of the intended landing site.

Surface mobility navigation capability

During six-month increments on the lunar surface, astronauts may embark upon excursions more than 500 kilometers from the outpost, a farside trek that will place the crew outside the communication range provided by DTE or LCT. To ensure crewmembers are able to find their way back to the Outpost, periodic fixes of land-marks, coupled with star tracking, will need to be taken *in situ* every few minutes, a process that will utilize one-way and two-way S-band Doppler information from the LRS (Table 5.9).

When embarking upon excursions closer to home, astronauts will utilize the LCT, a system that will have a 5.8-kilometer range and a 200 Mbps uplink capability to the LRS or to Earth. During these excursions, astronauts will also utilize cell phones and the lunar local area network (LLAN), which will support the portable fixed base radio (FBR) crewmembers will use when driving the Rover.

Table 5.9. Lunar relay satellite.

Description	Current best-estimate mass (kg)	Total mass (kg)	Nominal power (W)
Lunar relay satellite	1,033.5	1,124.2	683.6
Communications; two 100 Mbps high-rate links from surface	79.4	81.8	494
Avionics	91.6	113.3	189.6
Structures and mechanisms	180.5	205.6	0
Power System; 1,040 W average power load; two 1-axis solar arrays; 4.7^2 area	72.6	94.8	0
Propulsion; pressure-fed hydrazine	21.8	23.5	0
Propellant management	72.9	83.9	0
Propellant	467.3	467.3	0
Thermal control; Heat Pipe Radiator System; hydrazine heaters	47.3	54	0

Radiometric time architecture

Time will be a crucial factor during every lunar mission. To ensure astronauts are keeping to their schedules, time accuracy will be maintained by Mission Control on Earth. Mission Control will produce a navigation message transmitted to the LRS transceiver. The LRS transceiver will then compare that time against the atomic time and frequency standard, before transmitting the navigation message on a forward link to a surface-based transponder which will demodulate the navigation message and forward the message to crewmembers working on the lunar surface.

Surface communication systems

The lunar communications hub will provide astronauts with wired and wireless connectivity to major elements on the lunar surface such as other habitats and rovers embarked upon extended range or farside excursions. The hub will provide communications for line of sight and beyond line of sight via the LRS. It will also provide one-way and two-way ranging and Doppler tracking of elements deployed on the lunar surface. The hub's capability will extend to serving as many as 15 simultaneous users with an aggregated bandwidth of 80 Mbps at extended ranges of at least 5.6 kilometers. This capability will be achieved by using a suite of communication devices that will include fixed radios, mobile radios for rovers, and EVA suit radios. In the habitat, astronauts will have powerful 100 Mbps modulators available, used to downlink and uplink large data volumes pertaining to current operations.

Surface wireless mesh networking

With several astronauts working at the outpost, the communications traffic will be served by multiple wireless links operating simultaneously. Such a wireless environment may present communication challenges to astronauts while conducting EVAs on the surface due to the requirement of switching between different networks. On Earth, an automated mechanism for maintaining connection as one transits between networks is known as voice call continuity (VCC). To serve multi-link operations on the Moon, it is likely a system similar to VCC will be developed to ensure seamless handover from one element to the next.

SURFACE MOBILITY SYSTEMS

Lunar Surface Mobility Systems (LSMS) encompass both robotic rovers such as the Trencher Utility Rover (TUR) and Crawler Crane, and human rovers such as the Mobile Lunar Transporter and the All-Terrain-Hex-Legged Extra-Terrestrial Explorer (ATHLETE). This section describes the types of robotic and human rovers and examines the functions of each.

Robotic rovers

The lunar launch manifest recommends the landing of robotic rovers in late 2017. Once on the lunar surface, the rovers will be tasked with digging, carrying, grading, trenching, and a myriad of other construction-related activities dedicated to preparing for human exploration. To perform these activities, engineers designed three multi-purpose rover types capable of performing tasks ranging from breaking rocks to lifting massive items such as the Habitat Module. For example, the Backhoe Utility Rover has been designed with the primary function of placing regolith but it may also be used as a forklift to transport smaller elements such as airlocks. Similarly, the TUR has been designed with a high degree of multi-function capability, able to break up bedrock, dig trenches, and lay cable, whereas the Crawler Crane, as the designation implies, was designed to lift items such as the Habitat Module and cargo pallets. Other rovers that will be deployed may include the Recon Rover (Figure 5.5), capable of scouting out locations for construction activities.

Human-controlled rovers

One of the first human-controlled rovers to arrive on the lunar surface may be the Mobile Lunar Transporter (Figure 5.6, see color section). Designed for short distances, the Mobile Lunar Transporter will be used primarily during the initial 14-day missions.

Figure 5.5. NASA's Recon Robots will be used to perform highly repetitive and long-duration tasks, such as site mapping and science reconnaissance that would be unproductive for a human crew to conduct manually. Ames Research Center. Image courtesy: NASA.

Once exploration missions commence, the crew will be venturing farther from the base and will need to use a pressurized platform such as the ATHLETE configuration (Figure 5.7, see color section). The ATHLETE's multi-wheeled dexterity will allow astronauts to load, transport, manipulate, and deposit payloads to any site on the lunar surface, capabilities that are sure to make it the RV of choice.

The outpost design and development considerations described in this chapter are still in the process of being defined, evaluated, and optimized with regard to performance, efficiency, reliability, weight, volume, and cost-effectiveness. An effective outpost design will reduce system mass, reliance on ground controllers, and crew time spent monitoring systems. To achieve the optimal outpost design, NASA and the agency's centers, tasked with developing outpost technologies, continue to evaluate design criteria, resolve the challenges posed by radiation, pressurization, and human health, and provide solutions to key issues such as life support and power generation.

REFERENCES

[1] Anderson, B.J. *Natural Orbital Environment Guidelines for Use in Aerospace Vehicle Development*, NASA Technical Memorandum 4527. NASA George C. Marshall Space Flight Center, Huntsville, AL (1994).

[2] Bayless, D.; Brown, I.; Jones, J.A.; Karakis, S.; Karpov, L.; McKay, D.S. Novel concept for LSS based on advanced microalgal biotechnologies. *HABITATION 2006: Conference on Habitation Research and Technology Development* (2006).

[3] Benaroya, H. An overview of lunar base structures: Past and future. *AIAA Space Architecture Symposium, Reston, VA*, pp. 1–12. American Institute of Aeronautics and Astronautics, Washington, D.C. (2002).

[4] Bodkin, D.K.; Escalera, P.; Bocam, K.J. A human lunar surface base and infrastructure solution. *Space 2006, September 19–21, 2006, San Jose, California*, AIAA 2006-7336. American Institute of Aeronautics and Astronautics, Washington, D.C.

[5] Churchill, S. *Fundamentals of Space Life Sciences*, pp. 13–16. Krieger, Malabar, FL (1997).

[6] Eckart, P. *Life Support and Biospherics*. Herbert Utz, Munich, Germany (1994).

[7] NASA. *First Lunar Outpost Study*, NASA Working Group Report. NASA Ames Research Center, Moffett Field, CA (March 1992).

[8] Gugliotta, G. *U.S. Planning Base on Moon to Prepare for Trip to Mars*. The Washington Post, Washington, D.C. (2006).

[9] Heiken, G.; Vaniman, D.; French, B. *Lunar Sourcebook: A User's Guide to the Moon*. Cambridge University Press, Cambridge, MA (1991).

[10] Kliss, M. *Life Support Systems for Human Space Exploration beyond Low Earth Orbit*, Stanford University Lecture Series, Stanford University, Palo Alto, CA (March 7, 2006).

[11] Koelle, H.H. *Environmental Control and Life Support Systems (ECLSS) for MOONBASE 2015*. Technische Universität Berlin Institut für Raumfahrt (2000).

[12] Kubicek, K.; Woolford, B. *Man–Systems Integration Standards*, NASA-STD-3000. NASA, Washington, D.C. (1995).

[13] Larson, W.J.; Pranke, L.K. *Human Spaceflight Mission Analysis and Design*. McGraw-Hill, Columbus, OH.

[14] NASA. *Guidelines and Capabilities for Designing Human Missions*. NASA, Washington, D.C. (2003).

[15] Simonsen, L.C. Analysis of lunar and Mars habitation modules for the Space Exploration Initiative (SEI). In: J.W. Wilson, J. Miller, A. Konradi, F.A. Cucinotta (eds.), *Shielding Strategies for Human Space Exploration*, NASA Conference Publication 3360. NASA, Washington, D.C. (December 1997).

[16] Skoog, I. *Life Support Systems for Man: Life Sciences Research in Space*, ESA-SP-1105, pp. 97–108. ESA, Noordwijk, The Netherlands (1989).

[17] Spudis, P.D. *The Once and Future Moon*, pp. 83–101, 255. Smithsonian Institution Press, Washington, D.C. (1996).

[18] Taylor, L.A.; Schmitt, H.H.; Carrier, W.D.; Nakagawa, M. *The Lunar Dust Problem: From Liability to Asset*. American Institute of Aeronautics and Astronautics, Washington, D.C.

[19] Wieland, P.O. *Designing for Human Presence in Space: An Introduction to Environmental Control and Life Support Systems*. NASA George C. Marshall Space Flight Center, Huntsville, AL (1994).

6

Astronaut selection and medical requirements

A human being should be able to change a diaper, plan an invasion, butcher a hog,
conn a ship, design a building, write a sonnet, balance accounts, build a wall,
set a bone, comfort the dying, take orders, cooperate, act alone, solve equations,
analyze a new problem, pitch manure, program a computer, cook a tasty meal,
fight efficiently, die gallantly. Specialization is for insects.

Robert A. Heinlein.

In September 2007, NASA opened its applications for the 2009 Astronaut Candidate Class. Graduates of the Class of 2009 will not only have the opportunity to fly the agency's new Orion Crew Exploration Vehicle (CEV), the successor to the Space Shuttle, but also be eligible for missions to the Moon. Successful applicants will possess the standard shopping list of astronaut qualifications that include a PhD, extensive flying experience, scuba certification, and parachuting skills. Those selected will report to Johnson Space Center (JSC) in the summer of 2009 to begin the basic astronaut training program preparing them for future flight assignments which may include missions to the Moon.

Exploration-class missions to the Moon will demand unique selection criteria and mission-specific training but perhaps the most daunting challenge will be providing health and medical care to astronauts. In discussing each of these issues it is necessary to first understand the human element as it pertains to remote missions conducted beyond low Earth orbit (LEO).

PSYCHOLOGICAL IMPACT OF LONG-DURATION MISSIONS

Increasing space mission duration has presented a range of biomedical and behavioral challenges, most of these having been met with success under adverse circumstances. The human element, however, remains arguably the most complex component mission planners must consider when designing long-duration missions to the Moon. In fact, human interactions in an outpost isolated from Earth may represent one of the most serious challenges of exploratory missions.

Psychosocial issues

"All the necessary conditions to perpetrate a murder are met by locking two men in a cabin of 18 by 20 feet for two months."

Cosmonaut Valery Ryumin

Extended-duration spaceflight provokes a constellation of symptoms. These symptoms range from negative psychological problems such as sleep disturbances, mood disorders, and reduced energy levels, to interpersonal issues such as increased interpersonal tension and decreased team cohesiveness [14, 15]. Fortunately, thanks to the extraordinary motivation and professionalism of astronauts, none of these problems has had a negative impact upon a mission. However, there is anecdotal evidence the Russians once launched a rescue mission to the Mir space station for the purpose of returning one stress-stricken cosmonaut to Earth [4]. Nevertheless, although astronauts represent the most highly trained and motivated workers on and off the Earth, the fact remains even this select group have a threshold beyond which things may get out of hand.

Long-duration spaceflight in common with Antarctic and Arctic exploration expeditions require individuals to establish and maintain stable interactions between one another for prolonged periods in confined and isolated conditions. Such circumstances inevitably provide a breeding ground for interpersonal conflict. One of the problems with obtaining accurate data concerning psychosocial factors exhibited by astronauts is so few humans have flown in space. Of those who have flown in space, only a small percentage have been members of long-duration crews.

Analogs

To investigate the difficulties of living for extended periods of time in microgravity, psychologists study analogs of spaceflight such as Antarctic stations and nuclear submarines. Typical findings of those researchers who studied crewmen "wintering over" for eight months in Antarctic stations revealed a pronounced increase in stress-related symptoms of anxiety, depression, insomnia, and hostility [19]. However, the Antarctic crewmen studied lived in luxurious conditions compared with the cramped confines on-orbit. In space, astronauts are subject to limited privacy, noisy working conditions, time-consuming exercise countermeasures, and limited bathing facilities, usually consisting of a weekly shower with a sponge.

As NASA prepares to send its astronauts on long missions to the Moon, psychologists warn the potential for psychosocial problems may increase. This will be due in part to the increasingly precarious nature of the missions during which astronauts will be subject to the added risks of prolonged exposure to radiation.

Compounding the situation will be the aforementioned challenges of living in isolated and hazardous conditions for long periods.

To understand what sort of behavioral problems may be exhibited by long-stay lunar astronauts, researchers turn to the aforementioned analogs. However, these analogs are not perfect simulations of spaceflight since crew characteristics, screening procedures, mission objectives, and duration are different for each analog environment. Nevertheless, analog environments are the only tool researchers have to study the behavioral impacts of confinement, isolation, and prolonged periods of stress.

Perhaps the most common analog studies are those conducted on nuclear submarine crews (Figure 6.1) who stay underwater for up to six months. Since nuclear submariners are among the most highly trained, tested, and thoroughly screened individuals on Earth, it is not surprising the incidence of debilitative psychiatric illnesses among crewmembers is relatively low compared with the general population. The most common psychiatric symptoms reported by this elite group of undersea workers include anxiety, depression, and interpersonal problems [28].

Although the nuclear submarine environment replicates to some degree the confines of a spacecraft, many researchers agree the most useful analog is the Antarctic research station. This is due to the extreme environment, the extended tours of duty, and the limited contact with the outside world. Other features Antarctic research stations share with orbiting spacecraft include heterogeneity of

Figure 6.1. The environment onboard a nuclear submarine such as this Virginia-class vessel, imposes similar stresses to those experienced by long-duration space crews. Image courtesy: Northrop Grumman.

crewmembers, high skill levels, organizational similarities, and the rotational structure of tours of duty [19, 22, 26]. In fact, the conditions in the Antarctic and onboard orbiting space stations are considered so similar that Antarctica has served as a primary source of psycho-sociological data for predicting behavior onboard the International Space Station (ISS) and future interplanetary missions.

Ground-based space simulators

In 1990 the European Space Agency (ESA) locked six civilians with science and engineering backgrounds inside a hyperbaric chamber for four weeks as part of a simulation experiment designed to replicate the daily activities of astronauts.

The Isolation Study for European Manned Space Infrastructure (ISEMSI), as it was known, subjected the volunteers to various psycho-social tests designed to identify social and emotional conflicts and to record and analyse patterns of communication between crewmembers. The researchers observed group communication was balanced during the first few days of the experiment, but then communication patterns gradually changed. Towards the end of the confinement much of the communication was limited to two-way exchanges between crewmembers and the commander, although the group remained a cohesive unit throughout their stay.

A less successful study, which resulted in fistfights and charges of sexual harassment, took place during the Simulation of Flight of International Crew on Space Station (SFINCSS) experiment conducted by the Russian Institute for the Study of Biomedical Problems in 1999. During a party celebrating New Year's Eve, in the course of the 110-day isolation experiment inside a Mir simulator, Canadian Judith Lapierre was pulled out of view of the module's observation cameras and forcibly kissed by a Russian crewmember. The study had been designed to evaluate and observe social interactions of mixed gender crew in which crewmembers were housed in separate modules and executed different flight programs and, until the kissing incident, had largely gone unreported by the media. Following the fistfight and sexual harassment incident, conditions deteriorated to the point where the commander requested two crewmembers either be withdrawn from the study or have a hatch closed between two chambers to prevent further interaction.

Whether it be ISEMSI, SFINCSS, or ESA's upcoming 500-day Mars500 simulation scheduled for 2008, the results of this type of study typically reveal a distinct *us vs. them* syndrome, a feature previously documented in spaceflight and Antarctic stations. The syndrome is caused by the team perceiving that Mission Control does not understand the true nature of the problems faced in the isolated environment and therefore view the central authority as an outside authority, which inevitably sets up the potential for conflict.

Psychological problems during space missions

Perhaps the most infamous example of disagreement between a crew and Mission Control occurred during the November 1973 Skylab 4 mission, comprised entirely of rookie astronauts who complained their workload was excessive. Unsurprisingly, given the Skylab 4 astronauts had been given a similar workload to previous Skylab crews, Mission Control was unsympathetic to the complaints of Commander Gerald Carr, Pilot William Pogue, and Science Pilot Edward Gibson, and told the crew to continue with the assigned work. The astronauts, unhappy with the attitude of Mission Control, switched off the radio and staged a mutiny by declaring an unscheduled rest day. This act of insubordination resulted in a NASA rule stating at least one ISS crewmember must now have previous spaceflight experience. Examples of similar problems a crew may be subjected to are depicted in Table 6.1, which describes a sample of events that occurred during the NASA–Mir era.

Table 6.1. Events leading to adverse psychological effects during the NASA–Mir missions.[a]

Mission	Event	Details
Apr 95	Death of family member	Mission commander Vladimir Dezhurov learned of his mother's death
Feb 97	Fire and decompression	A fire in an oxygen-generating device began in the Kvant module of Mir forcing the crew to don breathing masks; the fire lasted 15 minutes
Mar 97	Oxygen generator failure	An oxygen generator fails, leaving the three-man crew with a two-month supply of oxygen
Apr 97	Ethylene glycol leak	NASA astronaut Jerry Linenger informs Mission Control ethylene glycol fumes leaking from the coolant system are giving the crew nasal congestion
Jun 97	Decompression	A Progress M-34 supply spacecraft was under manual control from Mir when rate-of-closure control was lost and the Progress M-34 collided with the Spektr module of Mir causing decompression of the Space Station
Mar 98	Computer failure	The space station's main computer, which controlled orbital alignment, failed, leaving Mir adrift until the crew managed to restart it.
Mar 98	Excessive workload	Commander Talgut Musabayev informed Russian Mission Control the crew is overworked and are making mistakes due to lack of rest
Jun 98	Air Conditioning System failure	Temperatures rose to 35°C; despite installation of a new air conditioner, temperatures remained at 28°C

[a] Adapted from Ark and Curtis (1999) [1].

CREW SELECTION CRITERIA

Medical selection of astronauts

The purpose of medical standards for exploration-class missions is ensuring astronauts are physically and temperamentally fit for the performance of orbital activities and extended operations on the lunar surface.

Defects and diseases will disqualify potential astronauts for a number of reasons and astronaut medical certification will only be issued after an individual has been assessed in each of the physiological systems and conditions listed below.

1. Endocrine system
2. Genitourinary system
3. Respiratory system
4. Cardiovascular system
5. Gastrointestinal system
6. Neurological
7. Psychological and psychiatric evaluation
8. Ophthalmology
9. Ear, nose, throat, and equilibrium
10. Musculoskeletal
11. Hematological and immunologic
12. Radiation exposure
13. General medical condition

For each physiological system, potential astronauts will be required to be free from any system-specific disorder which accredited medical conclusion indicates would render the crewmember unable to perform the duties required of an astronaut in training or during flight.

For each physiological system there are various disqualifying conditions. Some of these select-out conditions are discussed here.

Endocrine system

A potential disqualifying condition associated with the endocrine system is Type 1 diabetes mellitus, since an individual with this condition requires injections of exogenous insulin to properly metabolize carbohydrates and lipids [2]. Such a situation is clearly incompatible with spaceflight operations since, in the absence of treatment, a potential exists for disastrous incapacitation [2], which might jeopardize the individual and crew.

Genitourinary system

Several genitourinary conditions may result in an astronaut being medically disqualified due to the potential of such a condition to subtly or suddenly incapacitate, usually as a result of the severe pain that is often an accompanying symptom [17].

An individual who suffers from urinary tract calculi, for example, is also at a significant risk of infection with hematuria, frequency, and dysuria [17].

Respiratory system

A respiratory system and/or pulmonary function disorder increases an individual's susceptibility to dysfunction in the spacecraft environment as such a condition carries with it a potential for incapacitation, hypoxia, acceleration atelectasis, and compromise of g-tolerance. For example, an individual who suffers from chronic bronchitis and emphysema may experience significant hypoxia, a situation aggravated by the hypoxic and hypobaric environment of the spacecraft. Such an individual may also suffer from dysfunction of the bronchioles resulting in small airway occlusion in a high-g environment, such as experienced during launch. This problem may be exacerbated by the flight suit, which is fitted with g-protective equipment, resulting in a translocation of intrathoracic blood volume, a process that may further compromise small-airway function. Also, in the event of a rapid or explosive decompression (ebullism), individuals with weakened lung tissue will be at greater risk of pulmonary barotrauma.

Cardiovascular system

The majority of cardiovascular disorders are associated with the risk of sudden death or incapacitation [14, 29]. For example, coronary artery disease (CAD) is unpredictable and may be aggravated by circumstances such as heat, hypoxia and exposure to high $+g_z$, each of which increases myocardial oxygen demand. Another serious cardiovascular disorder is myocardial infarction, a condition associated with a range of pathophysiological behaviors of atheromatous plaques which have the potential to rupture and occlude vessels, a situation that has the risk of a potentially catastrophic and incapacitating event.

Gastrointestinal system

Several conditions of the gastrointestinal system may have implications for spaceflight, a number of which may be acute and/or chronic and vary in severity. For example, volumetric changes of intra-abdominal gases associated with a hypobaric environment or rapid/emergency decompression may precipitate an existing condition and may incapacitate an astronaut to such a degree that the astronaut's ability to perform an emergency egress is compromised.

Neurological

Several conditions, such as migraines, which may develop into migrainous stroke, have the potential to seriously compromise safety in the spacecraft environment. Similar risks exist for ischemic stroke, epilepsy, seizure, and syncope [17, 29].

Psychological and psychiatric evaluation

Anxiety disorders, mood disorders, and undesirable personality traits are some of the most common reasons for disqualification during the selection process of ESA, Canadian Space Agency (CSA), and NASA astronauts.

Any of these disorders are incompatible with spaceflight due to their potential to manifest themselves in overt acts such as panic and phobias [23]. Also, these disorders are unpredictable in their onset and may be triggered by a variety of stressors that have the potential to be disabling. The primary concern with personality disorders is the difficulty in effectively treating the individual, and since the disorders are usually so complex, making decisions about fitness to fly is so difficult as to be patho-gnomonic. Given the impact psychological traits may have upon a mission and crew, space agencies devote particular attention ensuring only the most psychologically suitable individuals are selected for spaceflight. This is often achieved by defining psychosocial/psychological select-out and select-in criteria based on clinical psychiatric evaluation and psychometric testing.

The "Wrong Stuff": select-out measures

The first step of a psychiatric selection strategy is to remove unsuitable candidates, a process achieved by identifying certain "select-out" disqualifying medical criteria such as schizophrenia and psychopathic deviation. To determine if an individual has a history of any disqualifying disorders, a formal clinical evaluation in the form of a psychiatric interview is performed. Other evaluation tools that may be applied include psychometric tests such as the Minnesota Multiphasic Personality Inventory (MMPI), the Rorschach Ink Blot test, and the Million Clinical Multiphasic Inventory (MCMI).

Psychiatric interviews are normally conducted with at least two independent psychiatrists. The interviews are carefully structured to ensure candidates provide as much clinical information as possible regarding a particular subject and are presented in a format designed to reduce the number of "no" and "yes" responses. For example, rather than simply asking "Have you ever been depressed?" to which most candidates will respond in the negative, the question is phrased "Tell me about the time where you have been the most sad in your life," thereby ensuring the candidate provides clinical information regarding the subject of depression [24].

The "Right Stuff": select-in measures

Spaceflight select-in measures are those traits in an individual identified as being desirable in an astronaut. Such criteria may include personality characteristics such as ability to tolerate stress, high motivation, and the ability to form stable inter-personal relationships. At the beginning of the manned spaceflight era these select-in criteria were driven by the characteristics of the type of mission and were limited to high stress tolerance, an ability to make decisions under extreme pressure, and exemplary piloting skills. However, with the advent of the Space Shuttle era, astronauts were required to possess engineering, scientific, and medical experience, which required a re-evaluation of the validation of select-in criteria.

Psychological profile of astronauts

The use of personality assessment is recognized as an important approach by space agencies for selecting candidates likely to perform and adapt optimally to space. However, no standardized psychological test for selecting astronauts exists. Research investigating the role of personality characteristics has revealed trait assessments do not play a major role in final-stage astronaut selection. Furthermore, no discernible differences in psychological testing between applicants eventually selected and those who were not have been identified. Nevertheless, given the recent media attention in 2007 devoted to the case of NASA astronaut Lisa Nowak, who was accused of trying to kidnap a romantic rival, it is likely psychological screening of some sort will play a role in selecting those astronauts who will return to the Moon.

Ophthalmology

The space environment is a hostile one, characterized by zero-gravity, hypoxia, possible ebullism, high-speed acceleration, and electromagnetic glare, each of which has the potential to degrade vision and, in individuals with certain disorders, the potential to blind, disable, and disorient. Each crewmember will be required to perform tasks requiring adequate depth perception, color vision, and spatial discrimination. A deficiency in any one of these ophthalmologic functions may jeopardize other crewmembers in the event of an emergency and, in the event of an emergency decompression, those predisposed to visual defects may suffer transitory hemiplegia, monoplegia and, in rare cases, permanent visual impairment.

Ear, nose, throat, and equilibrium

Hearing, balance, speech and communication, and unrestricted breathing are especially important to the safe conduct of spaceflight operations. Certain disorders, such as vertigo, may present significant risk to other crewmembers as the condition may be associated with intractable symptoms that include episodic vertigo, fluctuating hearing loss, and ear pressure. Another serious disorder is benign positional vertigo, a condition occurring without warning and associated with vertigo, nausea, and vomiting. Needless to say, an individual exhibiting such symptomatology, which would be greatly aggravated in microgravity, would present a serious risk to other crewmembers.

Musculoskeletal

Since each crewmember must be able to perform mission-critical procedures in the event of an emergency and to conduct the necessary procedures to perform an emergency egress, certain musculoskeletal conditions associated with physical handicaps are considered disqualifying.

Hematological and immunologic

Specific control measures are implemented by NASA such as a limited access/ quarantine program prior to launch to prevent exposure to infectious disease during the preflight period. In addition to prevention, there are certain diseases that will disqualify the candidate astronaut due to the requirement for ongoing administration of medication. HIV, for example, is an infection associated with symptomatology, treatment, and potential for side-effects clearly incompatible with spaceflight.

Radiation exposure

The issue of radiation exposure was mentioned briefly in the preceding chapter. Crewmembers embarked upon long-duration missions may approach or exceed career radiation limits due to the harsh radiation environment of the lunar surface. To mitigate the risk of radiation exposure, NASA may select crews who have low career radiation levels or the agency may, if screening technology exists by the time lunar missions commence, select those astronauts who are less susceptible to exposure to ionizing radiation.

Crew compatibility

Crew selection has often been viewed as an opaque process, in terms of the individuals selected and crew composition. Perhaps the most diverse crew ever launched into orbit was STS-51G, comprising civilian and military NASA astronauts, both male and female, a Saudi Arabian prince, and a French cosmonaut. Despite the obvious multinational nature of the mission and the obvious cross-cultural challenges the crew of STS-51G was, by all accounts, a harmonious one that was compatible and worked effectively throughout the mission. However, typical Space Shuttle missions last no more than two weeks, whereas an exploration-class mission will last as long as six months. Crew compatibility issues therefore will obviously assume increasing significance in determining the effectiveness of the mission.

The problem in determining crew compatibility comes down to there being no one measure to predict if a crew will work together effectively. Some researchers favor the use of psychological performance tests and personality questionnaires whereas other investigators prefer a more behavior-oriented approach. The Russians, on the other hand, who have invested considerably in developing methods to assess interpersonal compatibility, consider biorhythms to be a useful tool for selecting cosmonauts.

Crew composition

NASA plans to send a crew of four to the Moon, a number that fits with the current belief a crew should be as small as possible, since the larger the crew the greater the potential for interpersonal conflict. The occupational role of each member of the lunar crew has yet to be determined, but it is almost certain one crewmember will be a pilot and it is likely another crewmember will be a medical doctor, given the extended duration of the mission. The role of commander will be assigned to the crewmember with the most experience and will not necessarily be the pilot, as has been the case in so many space missions.

Leadership

Fewer than one hundred astronauts have commanded a space mission. Given the dynamic and challenging environment of the lunar surface, combined with the difficulties posed by six-month missions, commanders of lunar missions will be required to perform to the very highest standards.

Fortunately, unlike many of the other aspects of returning to the Moon that require research, there is no requirement to conduct studies to characterize the qualities constituting a prerequisite for effective leadership in an exploration-class mission. Those interested in the qualities most suited to leading small groups for several months in a hostile environment have a wealth of information from the days of polar exploration. The expedition accounts of legendary explorers such as Fridtjof Nansen, Roald Amundsen, and Douglas Mawson provide a plethora of valuable lessons for those destined to lead lunar missions. However, of all the great explorers of the early 20th century, one man stands above the rest. His name is Ernest Shackleton (Figure 6.2, see color section) and he has been called "the greatest leader that ever came on God's earth bar none" after saving the lives of his crew who were stranded with him in the Antarctic for almost two years. To appreciate the amazing leadership skills that classed Shackleton as a true master of men and to understand why those skills will be as effective on the lunar surface as they were in the Antarctic nearly one hundred years ago, it is necessary to revisit Shackleton's incredible tale of survival.

Shackleton's expedition was to be the first to cross Antarctica. In August 1914, he and his crew set sail aboard the *Endurance* but in the dangerous waters of the Weddell Sea, Shackleton's ship became trapped in the ice pack. For the next ten months Shackleton and his crew waited for the ice to break but, instead, the ice crushed the *Endurance*, leaving the men shipwrecked almost 2,000 kilometers from civilization.

Stranded on the ice, Shackleton and his crew suffered in appalling conditions as the temperature plummeted. Temperatures became so cold clothes froze and many of the crew suffered frostbite, necessitating operations to cut off crewmembers' toes. Subsisting on a daily diet of penguin, Shackleton and his crew suffered through the endless night of the Antarctic winter, becoming progressively weaker and malnourished. When the ice finally broke they were forced to make their escape in three small lifeboats and survive on the sea, which they did for four months. Storms and 15-meter waves were daily events as Shackleton and his crew made their way to Elephant Island, only to find it inhospitable.

The next several months were spent living under the carcasses of two upturned lifeboats, until Shackleton decided he would take five men in one of the patched-up lifeboats in a desperate attempt to reach the inhabited island of South Georgia. Thanks to the greatest navigation exploit in seafaring history, Shackleton and his crew of five steered across more than 1,200 kilometers of the most savage ocean in the world to reach South Georgia. On reaching the shores of South Georgia they discovered they had to cross a mountain range to reach the whaling station. Somehow, Shackleton and his men managed that amazing feat, after which Shackleton returned to Elephant Island and rescued the remainder of his crew. Incredibly, every man survived.

How did Shackleton do it? First, to lead his men in a crisis situation Shackleton understood the most effective way to get them to follow was to lead by example. A case in point exemplifying this type of leadership was when Shackleton and his men arrived at Elephant Island at three in the morning. Every man was in a state of complete exhaustion, including Shackleton, but it was he who took it upon himself to take the first watch. By doing that, he built a strong team who were happy to put their trust in him. Second, during the long months stranded on the ice, Shackleton mixed with the crew and the officers and took little notice of the hierarchy conventional at the time.

Unsurprisingly, Shackleton's leadership model is used extensively today by organizations ranging from the Special Air Service to international business corporations. The successful lunar commander will be the one who adopts the principles of Shackleton's management style. He or she will understand everyone shares in the responsibility for the mission's success but the final decision on which course of action to take will rest with the commander. The lunar commander will also possess a great sense of optimism, enabling him/her to keep going in the toughest conditions while at the same time inspiring others to do the same. Shackleton once said: "Optimism is true moral courage," and in the event of an emergency on the lunar surface, such an outlook may well mean the difference between living and dying.

CREW TRAINING

For the first two years of their training astronauts selected for lunar and exploration-class missions will follow a program similar to that followed by Space Shuttle and ISS astronauts.

Basic training

Initially the Class of 2009 will be designated astronaut candidates (ascans) and be assigned to the Astronaut Office at the JSC. There, the ascans will undergo a training and evaluation period lasting nearly two years. During this training cycle they will participate in the basic astronaut candidate training program, designed to develop the knowledge and skills required for formal mission training. For example, during their training program each ascan will become scuba-qualified, complete the military water survival requirement, a swimming test, and complete various other training qualifications. At the end of their training, ascans will be declared trained astronauts and be eligible for mission training. At this juncture, for those chosen for an exploration-class mission, the training will differ somewhat from the typical preparation normally conducted by astronauts selected for ISS increments or Space Shuttle missions.

Lunar training

NASA has not explored another world since 1972. Therefore, to characterize the training the new crop of astronauts will need to undertake before traveling to the Moon it seems reasonable to assume some of the guiding principles and lessons learned from Apollo will be useful in the training of the next generation of lunar crews.

The Apollo astronauts performed extensive geologic field training, which proved to be invaluable to the achievement of scientific objectives such as sample acquisition and documentation. The new breed of lunar astronauts will also conduct geological field training exercises, many of which will take place in analog environments such as the Haughton Mars Crater, the Big Island of Hawaii, and Iceland. This training will not only serve to support the testing of equipment but also provide geoscientists with an opportunity to teach the new cadre of astronauts the finer points of lunar geology. The training will also prove invaluable for reducing operational risk and provide astronauts with the opportunity to acquire the skills and understanding necessary to make meaningful observations about the lunar surface.

Emergency training

During their training for exploration-class missions astronauts will become familiar with JSC's Building 9 which houses a practice arena in which astronauts perform emergency scenario training. The emergency scenarios training will train astronauts to respond to fire/smoke, rapid/explosive decompression, and toxic release events. In each of the scenarios time will be critical. For example, a micrometeorite punching a one-centimeter hole in the Surface Habitat may give astronauts only 30 minutes to respond before their air supply is vented to space as a result of rapid decompression of the habitat.

Virtual Environment Generator training

Astronauts will also be trained in the use of the Virtual Environment Generator (VEG). The VEG is a virtual reality (VR) system that can simulate certain aspects of microgravity, assist in navigating new environments, such as the Lunar Habitat, and serve as a countermeasure to spatial disorientation. The VEG comprises a head-mounted display (HMD), the position and orientation of which commands a computer to generate a scene that corresponds to the position and orientation of the operator's head. This synthetic presence permits the operator to move around in the artificial world of the Lunar Habitat or even traverse the lunar surface.

When the astronaut dons the VEG equipment, he/she will be presented with an image of the interior of the habitat and a space-stabilized virtual control panel with an image of the astronaut's hand in the HMD. As the astronaut moves their hand, the virtual hand will also move. Collision detection software in the graphics computer [5, 7, 21] detects when the operator's hand penetrates the virtual control panel, enabling the astronaut to interact with the virtual switches or objects to control events within the habitat.

Astronauts will also be able to manipulate objects in the virtual habitat and be able to experience resistance to movement, texture, mass, and compressibility thanks to the haptic (tactile) and force feedback [5, 7] systems. To help astronauts in the virtual habitat, the system has been designed to provide auditory cues when an object is grasped or dropped, or when a virtual switch is operated. This synthesis of visual and auditory cues will augment the visual information presented to the operator, thereby enhancing the performance of crewmembers within the habitat.

Software ensures database compression techniques result in the virtual habitat containing all objects one would expect to see in the real habitat. Software also takes into account the effect of human behavior and the effect of collision for real-time operation, which means no matter how fast the operator moves through the environment, he/she experiences no visual lags [25]. The real-time operation results in the operator experiencing the high degree of realism and interactivity necessary to allow crewmembers to perform tasks necessary for training.

Psychological training

Once a lunar crew has been selected, one aspect of their training will be directed at developing interpersonal dynamics not only between each other but also between the crew and ground control personnel. This type of training is already common in both the American and Russian space programs and it will become even more important as missions become longer and crews are exposed to the dangers of radiation and microgravity for extended periods.

To help develop their interpersonal dynamics, astronauts will be assisted by the NASA Psychological Services Group (PSG) which began in 1994 and is composed of a group of behavioral scientists and psychologists. It was designed to support American astronauts during their stays onboard Mir and will no doubt play an equally important role for astronauts embarking upon a six-month tour of duty on the surface of the Moon. In addition to helping astronauts adapt psychologically, the PSG will provide input on a number of mission variables, ranging from work and rest schedules to training crewmembers to recognize potential inflight psycho-sociological issues.

Another important support unit tasked to help the astronauts and their families will be the Family Support Office (FSO), a group including representatives from the PSG, the Astronaut Office, and the Astronaut Spouses Group. While astronauts are deployed to the lunar surface for six months at a time, the FSO will serve as a liaison with NASA, which will provide regular contact with family members and update information regarding the mission. For example, some of the events the FSO may be involved with include organizing family conferences on special occasions, compiling digital family picture albums, and debriefing the family following the mission.

MEDICAL CARE

Astronauts represent the most carefully screened and extensively trained humans on Earth. However, the medical challenges of living for six months in an environment as harsh and unforgiving as the Moon mean it is inevitable that incidents of illness and occasional medical emergency will occur. Furthermore, it is known that astronauts exposed to extended periods of microgravity suffer a variety of physio/psychological decrements resulting in operational constraints. These decrements often necessitate countermeasures and rehabilitation, but it has been documented that many of these countermeasures are not fully protective, consume valuable crew time, and occasionally compromise mission effectiveness. Consequently, the provision of adequate medical care for exploration-class missions to the Moon presents a unique challenge to mission planners since, unlike long-duration missions to the ISS, there is no rapid return option from the lunar surface.

Radiation

Radiation exposure represents the greatest risk to crewmembers working on the lunar surface. Moon-bound crewmembers will be exposed to high-energy heavy ions, proton and photon radiation, and unexpected SPEs and CMEs (Figure 6.3). While the fluence (number of particles) of heavy ions will be much less than that of protons, the energy deposition from individual particles will be much greater [11]. Also, since conventional shielding will not effectively protect the crew from cosmic radiation, astronauts will also be exposed to a combination of secondary particle radiation and highly penetrating neutrons [8]. Although these astronauts will be exposed to lower levels than a cancer patient undergoing chemotherapy, when the planned six-month duration of the mission is considered, it is possible crewmembers may be exposed to potentially health-threatening levels of radiation [11].

Radiation types

GCR represents the dominant source of radiation that must be dealt with by lunar inhabitants since they include heavy, high-energy ions of elements able to pass practically unimpeded through an astronaut's skin. A secondary threat is represented by SPEs consisting primarily of protons and alpha particles and a heavy iron component which varies from event to event. The source region is believed to be near active regions on the Sun's surface, with possible secondary acceleration associated with CMEs. Since SPEs are associated with active regions on the Sun, they are more frequent near solar maximum and a single active region may produce a few SPEs over a period of weeks.

Radiation damage

Radiation causes damage to human tissue by ionizing atoms in the tissue, a process with enough energy to remove electrons from the atoms that make up molecules in the tissue. When the electron shared by the two atoms to form a molecular bond is

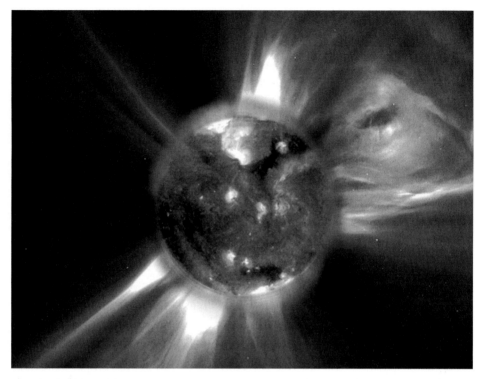

Figure 6.3. Coronal mass ejections have the potential to inflict serious radiation injury upon crewmembers embarked upon long-duration lunar missions. Image courtesy: NASA. Source: *http://antwrp.gsfc.nasa.gov.apod/ap070206.html*

dislodged by ionizing radiation, the bond is broken and thus the molecule falls apart. When ionizing radiation interacts with cells, it may or may not strike the chromosomes, considered the most critical part of the cell since they contain the genetic information and instructions required for the cell to perform its function and to self-replicate.

The least dangerous effect of radiation upon cells occurs when ionization forms chemically active substances which in some cases alter the structure of the cells. These alterations may be the same as those changes occurring naturally in the cell and may have no negative effect. A more damaging effect upon cells occurs when a radiation-damaged cell needs to perform a function before it has had time to repair itself. If this occurs, the cell will either be unable to perform the repair function or perform the function incorrectly or incompletely. The result may be cells that cannot perform their normal functions or that are now damaging to other cells. These altered cells may be unable to reproduce themselves or may reproduce at an uncontrolled rate. Such cells can be the underlying cause of cancer. At the most extreme end of the scale are cells dying as a result of the damage, which may depend on how sensitive the cells are to radiation. Cells which divide rapidly tend to show effects at lower doses of radiation than those which are less rapidly dividing. An example of a sensitive cell is the blood cell, part of the hemopoietic system which is the most sensitive biological indicator of radiation exposure.

Radiation-induced changes randomly distributed in the DNA of single cells may lead to cancer or genetically transmissible effects, depending on the target cells. These effects, known as *stochastic* effects, are the most important consideration in setting protection limits for human populations exposed to low-dose radiation, but at present there is no known safety threshold for stochastic effects.

Another damaging group of effects are known as *nonstochastic*. They may range from acute radiation sickness to hair loss or nausea and normally occur following exposure to relatively high doses of radiation. In contrast to stochastic effects, nonstochastic effects are dose-dependent in both frequency and severity and may occur early, in a matter of hours or days, or late, after many months. These effects are specific to each exposed individual and are characterized by a certain minimum dose that must be exceeded before the particular effect is observed. Because of this minimum dose, nonstochastic effects are also referred to as *threshold effects*, since the threshold may differ between individuals.

Acute radiation syndrome

The degree to which radiation inflicts damage is described by radiation dose measurements. The standard unit of radiation dose, the gray (Gy), is defined as a joule of energy absorbed as ionization and excitation per kilogram of tissue, delivered during a period of a few days. An acute radiation dose is defined as a large dose that exceeds 1 Gy. Such an exposure may cause a pattern of clearly identifiable symptoms and/or syndromes referred to as acute radiation syndrome (ARS). This syndrome is an acute illness caused by irradiation of the entire body or most of the body by a high dose of penetrating radiation in a very short period of time. Such an event might be experienced by astronauts during an SPE or a "worst-case scenario" solar flare. ARS results in three distinct syndromes:

- *Bone marrow syndrome* occurs with exposures between 0.7 Gy and 10 Gy and is characterized by damage to cells dividing at the most rapid rate, such as bone marrow lymphatic tissue. Symptoms include internal bleeding, fatigue, and bacterial infections and the survival rate of patients affected by this syndrome decreases with increasing dose, the primary cause of death being the destruction of bone marrow.
- *Gastrointestinal syndrome* occurs following exposures between 10 Gy and 100 Gy and is characterized by damage to cells dividing less rapidly, such as the linings of the stomach. Symptoms include nausea, vomiting, electrolytic imbalance, loss of digestion ability, and bleeding. Due to the destructive and often irreparable changes associated with this syndrome, survival is extremely unlikely with death usually occurring within two weeks.
- *Cardiovascular/central nervous system (CNS) syndrome* occurs following exposure to greater than 50 Gy and is characterized by damage to cells that

do not reproduce, such as nerve cells. Symptoms include loss of coordination, confusion, coma, shock, in addition to the symptoms of the blood marrow and gastrointestinal tract syndromes. Death normally occurs within three days due to collapse of the circulatory system.

Regardless of the type of syndrome, each is comprised of four stages, the first being the prodromal period, characterized by symptoms such as vomiting, nausea, and anorexia, the magnitude of which is dose-dependent. Symptoms may last minutes or several days. The latent period is characterized by the patient looking and feeling healthy for a few hours or even weeks, whereas the symptoms associated with the manifest illness period will depend on the specific syndrome and may last for hours or months. Finally, the patient either recovers or dies. Most patients who do not recover will die within months of exposure and those who do pull through may face a recovery period lasting for weeks or up to two years.

Two of the most serious consequences of ARS include loss of reduced fertility and transient fertility caused by gonadal irradiation. The nature of these effects and the doses required to produce them vary in males and females, but, typically, for the male the doses required to cause temporary sterility fall in the range of 0.5 Gy to 4.0 Gy for single acute exposures although a single acute dose as low as 0.15 Gy may produce a decrease in sperm count. The duration of temporary sterility is dose-dependent and may last from 8 to 10 months up to several years. Permanent sterility has been reported following doses in the range of 2.5 Gy to 4.0 Gy. Radiation doses necessary to sterilize most females fall in the range of 6.0 Gy to 20 Gy, although temporary sterility may occur at doses as low as 1.25 Gy and doses of 2 Gy to 6.5 Gy are required to sterilize 5% of women for more than 5 years.

Potentially late effects following radiation exposure during spaceflight include induction of cancer and damage to the CNS although, due to the problems quantifying these risks, there are many uncertainties with risk assessment. The risk of cancer is due in part to the effect of radiation upon DNA which is caused by the passage of an energetic charged particle through a cell, producing a region of dense ionization along its track. The ionization of water and other cell components may damage DNA molecules near the particle path but a "direct" effect breaks DNA strands. Single-strand breaks (SSBs) are fairly common and double-strand breaks (DSBs) less so, but both can be repaired by in-built cell mechanisms. Clustered DNA damage in which both SSBs and DSBs occur may lead to cell death, although this is not usually a problem since cells are replaced by normal processes. The more dangerous event is the non-lethal change in DNA molecules which may lead to cell proliferation, a form of cancer. This is because an attempt to repair a DSB may create one or more point mutations or it may induce the movement or duplication of a larger section of the DNA between chromosomes or between different regions of the same chromosome.

Radiation limits

Current radiation exposure limits for LEO are based on keeping exposure as low as reasonably achievable (the ALARA principle) and are contained within the National

Council on Radiation Protection and Measurements (NCRP) Report No. 98, which states the career limits for astronauts. However, these limits are based on equivalent doses to blood-forming organs (BFOs) and not on the effective dose to the whole body, so these guidelines cannot be used for estimating radiation exposure limits for lunar astronauts since the high-energy radiation from GCRs and SPEs results in whole-body exposure. Despite this, current research favors the use of LEO limits as guidelines for exploration-class missions. Such a policy may result in the computation of politically acceptable levels, but in reality may mean a death sentence to astronauts since the basis for radiation damage to mammalian cellular systems in LEO is very different from deep space. For example, continuous low dose rate heavy-ion radiation from GCRs results in cell damage that is highly variable for different cell types and is very difficult to calculate with little ambiguity. Also, calculating the hazards from exposure to ionizing radiation remains an unresolved issue, and there remains a significant shortage of data concerning biological response to GCR exposures.

Before describing some of the strategies NASA may implement to protect crews from radiation exposure, it is useful to compare radiation doses for different mission profiles. Unfortunately for those tasked with determining what represents a dangerous radiation exposure, information concerning deep-space radiation is lacking. Although a significant amount of data has been collated on radiation exposure during spaceflight (Table 6.2), much of this has been obtained from missions in LEO and it is difficult to extrapolate this information to deep space.

Current experience in long-duration missions is restricted to increments onboard the ISS, where, during a standard six-month tour of duty, an astronaut may expect to receive a dose of 0.5 Sv to 0.6 Sv per year. This amount is equivalent to ten times the maximum permitted dose for an employee working at a nuclear power plant! The medical effects of this long-term exposure to radiation may be manifested as uncontrolled cell division, DNA fragmentation, and ultimately apoptosis (cell death). Equally dangerous to lunar surface-bound astronauts is the threat posed by large

Table 6.2. Radiation doses during space missions [8, 11].

Mission	Absorbed dose (mGy)[a]	Dose equivalent (Sv)[b]
7-day Space Shuttle mission (orbit altitude <450 km)	2–4	0.005
8-day Space Shuttle mission (orbit altitude >450 km)	5.2	0.05
Apollo 14 (9-day mission to the Moon)	11.4	0.03
Skylab 4 (84-day mission; orbit altitude @ 430 km)	77.4	0.178
Mir (1-year mission; orbit altitude @ 400 km)	146	0.584

[a] One gray (Gy) equals the amount of ionizing radiation corresponding to 1 joule absorbed by 1 kilogram of material. By comparison, one chest X-ray is equal to approximately 1 mGy.
[b] The sievert (Sv) is the standard unit used for humans, where 1 Sv is equal to 100 rem (radiation equivalent man).

Table 6.3. Astronaut radiation exposure limits (Sv).

Period	Blood-forming organs	Eye	Skin depth
30 days	0.25	1	1.5
Annual	0.5	2	3
Career	1–4[a]	4	6

[a] Based on a maximum 3% lifetime risk of cancer mortality.

SPEs which may inflict high dosages of radiation in a very short period of time and cause pronounced *nonstochastic* symptoms and eventually ARS.

Obviously, given the potential for radiation to inflict such serious biological damage, it is important to establish threshold exposure limits, a calculation based on theoretical models considering the interaction of heavy particles with biological tissue. Unfortunately, the mechanisms by which biological tissue is damaged by radiation are still poorly understood and the database of radiation effects as a result of space radiation is very small. Consequently, the estimate of carcinogenic risk set for astronauts by NASA represents an approximation at best (Table 6.3).

Risk mitigation for acute radiation exposure

Several strategies have addressed risk mitigation of acute radiation exposure. For example, it has been suggested that only those astronauts with low career radiation levels be selected for long-duration Moon missions. Another suggestion is declaring astronauts with high radiation susceptibility ineligible for long-duration missions. This proposal is based on the fact that certain individuals possess certain genotypes making them more susceptible to adverse health effects associated with exposure to ionizing radiation. While the probability of an individual with these genotypes enrolling as an astronaut is low, there is still the potential for certain genotypes to confer an increased sensitivity to radiation. Unfortunately, it is not presently possible to screen individuals for radiation sensitivity based on DNA.

Given that at present there is no current viable primary prevention strategy for lunar astronauts, mission planners direct their focus upon secondary prevention game plans. These include provision of effective shielding, safe-haven modules, and the provision of radioprotective compounds.

Radioprotective agents

Some symptoms of radiation damage may be prevented medically through the administration of *amifostine* which protects tissues against ionizing radiation damage [12]. Another possible drug radioprotective agent, which reduces the deterministic effects associated with radiation exposure, is WR-2721. WR-2721 is administered with its active metabolite, WR-1065, and together these drugs may be administered up to 3 hours post-exposure. However, although the WR-2721/WR-1065 combination is capable of reducing the mutagenic and carcinogenic effects of radiation, it cannot be taken daily due to the toxic effects of even low daily doses.

Nanoparticles

Another secondary prevention strategy is the accurate monitoring of the biological impact of radiation using nanoparticles. Lunar mission planners may require astronauts to be injected with a clear fluid containing nanoparticles and be fitted with a small monitoring device in their ears. The fluid will contain millions of nanoparticles serving as a genetic platform to detect signs of radiation damage. The monitoring device will use a laser to count radiation-infected cells as they flow through the capillaries in the eardrum. A wireless link from the device will send information to the primary computer, which will then calculate the extent of the radiation damage and relay this information to mission control for evaluation.

Mission planning

Based on studies conducted by the NCRP, the optimal time for a lunar mission would be during solar minimum. During solar minimum there is a reduction in SPE occurrences and a corresponding increase in the GCR fluence rates, whereas during solar maximum there are increased SPE occurrences and a reduced GCR fluence rate. However, the purpose of constructing an outpost is to establish a permanent human presence on the Moon, so it will be necessary to remain on the Moon during periods of solar minimum and solar maximum.

Radiation monitoring

Another preventive measure NASA will implement during Moon missions is aggressive monitoring of space radiation using passive dosimeters and other radiation-monitoring tools such as the Tissue Equivalent Proportional Counter (TEPC), Charged Particle Directional Spectrometer (CPDS), and the Real-Time Radiation Monitoring Device.

Radiation Warning System

NASA will also use solar observatory satellites such as the Solar and Heliospheric Observatory (SOHO) as monitoring tools to help warn crew of a solar flare or radiation-laden cosmic winds.

Shielding

Although pharmacologic prevention and warning systems represent valuable preventive measures, the most effective means of protecting the crew is by employing adequate shielding. Although most solar flare energy is composed of alpha particles and beta particles that can be stopped by a few centimeters of shielding, the heavy, slow-moving, and damaging nature of cosmic rays requires shielding several meters thick to prevent penetration.

Table 6.4. Classification of stressors in long-duration spaceflight [1, 26].

Physiological	Psychological	Psychosocial	Human factors	Habitability
Radiation	Isolation	Crew coordination demands	High and low workload levels	Confined space
Altered circadian rhythms	Limited rescue capability	Interpersonal tension between crew and control	Limited exchange of communications	Sleep disruption
Sleep disturbance	High-risk environment	Disruption of family life	Mission danger and risk of equipment failure	Lack of privacy
Decreased exposure to sunlight	Alterations in sensory stimuli	Multicultural issues	Adaptation to artificially engineered environment	Isolation from support systems

Human behavior and performance

There will be a number of behavior-shaping constraints in the habitat and lunar environment with the potential to negatively impact behavior and compromise crew performance. The hazards of the high-radiation environment combined with the isolation, artificial environment, and the potential for unanticipated events will collectively conspire to pose a significant challenge to the cohesive functioning of the crew (Table 6.4). Although some of the problems associated with the behavior-shaping constraints will be reduced by the aforementioned careful selection and training of crew, there are ways habitat designers can help lunar inhabitants cope with the psychological stresses of isolation and mission challenges. For example, interface design will need to provide adequate information to enable informed decision making between crewmembers and Earth, as well as being able to support the actions and planning of the crew during nominal and crisis scenarios. In addition, interface design must be planned to reduce crew cognitive workload and fatigue in order to reduce human errors. Habitat designers can also help crewmembers cope by adding features similar to terrestrial working environments such as a small library and recreation area, as well as incorporating familiar visual cues, color, and light into the design.

Deconditioning

The reduced gravity environment of the Moon will cause the cardiovascular and skeletal system of crewmembers to undergo a process of deconditioning which involves adaptive changes in both structure and function. The severity of cardio-vascular and skeletal deconditioning is known to increase with spaceflight duration and may causes cardiac dysrhythmias, syncope [18], diminished cardiac function, and depleted bone density upon return to Earth.

Bone loss

Living on the Moon will cause a significant reduction in the mechanical load on the skeletal system, placing astronauts at risk of disuse osteoporosis. In fact, so deleter-ious is the effect of long-duration spaceflight upon the skeleton, some astronauts who spent six months onboard Mir more than ten years ago have yet to recover their bone mass. For this reason, many space medicine experts consider bone loss to represent the most serious physiological hazard associated with exploration-class missions.

The process by which the body divests itself of bone mass is not fully understood, but it is known that bone and muscle respond to reduced gravity within the first few days of a mission and the most severe loss of bone mass occurs between the second and fifth month in space [16]. Although lunar astronauts will not be expected to lose as much bone mass as ISS crews the process of demineralization is potentially more dangerous due to the requirement for surface excursions. This mission requirement will subject astronauts' weakened bones to undue stress, thereby placing crew-members at increased risk of a reduced fracture threshold in the event of a fall.

Microgravity-induced loss of bone mineral density (BMD) has been documented primarily in the weight-bearing components of the skeletal system such as the femur, pelvis, and spine. Several research studies have corroborated data indicating astronauts may lose between 1% and 2% of their BMD per month [16, 27], a rate approximately five times the rate of women with postmenopausal osteoporosis. Astronauts embarked upon a long-duration mission to the Moon may also, as a result of exposure to high levels of radiation, be susceptible to additional negative effects on bone health due to osteoradionecrosis (ORN). ORN is a condition of nonliving bone in a site of radiation injury [10] resulting in the denaturing of collagen fibers and has been observed in cancer patients receiving high doses of radiation during chemotherapy. Although research has not investigated the effect of ionizing radiation on general bone quality, there is a high risk that Moon-bound astronauts may be exposed to radiation sufficient to cause significant losses in both bone volume *and* bone integrity.

In addition to the increased risk of fracture as a result of osteoporosis, astronauts will also be susceptible to toxic accumulations of excess minerals in the kidneys and potentially irreversible damage to the skeleton.

Muscular deconditioning

The muscular system undergoes a similar deconditioning process although, unlike their ISS counterparts, astronauts on the Moon will be exercising the muscles used in standing, walking, and posture maintenance. However, the weak gravity of the Moon will still cause a gradual loss of strength and a reduction in size of many muscles.

Countermeasures

To offset the deleterious effects upon the cardiovascular, skeletal, and musculo-skeletal system, astronauts will spend at least two hours per day performing a broad program of exercise countermeasures designed to load the skeleton and induce mechanical strain upon the muscles.

Crewmembers will follow an individual exercise training program developed preflight, based upon exercise testing, prior flight experience, and level of conditioning. Following each exercise session, physical trainers will receive a downlinked file containing heart rate and ergometer data of the crewmembers' exercise training sessions. Every month the crew will be subject to fitness assessments to determine aerobic capacity and strength levels. Based on the fitness assessments, recommendations and changes will be made to the exercise program to ensure countermeasures continue to be effective.

In addition to the exercise equipment (Figure 6.4), the outpost will be provided with a suite of flight-proven non-exercise hardware such as a lower-body negative pressure (LBNP) device that simulates the gravitational force on the legs by drawing blood from the upper extremities.

Later in the outpost's development, it is possible that innovative countermeasure hardware, such as a centrifuge, will be installed in the health facility. Research has demonstrated centrifuge-induced artificial gravity to be effective in mitigating the

Figure 6.4. Lunar-bound crewmembers will be required to exercise between two and three hours per day to offset the deleterious effects of reduced gravity. Image courtesy: NASA.

negative effects of reduced gravity [14], and the installation of a human centrifuge on the Moon will provide an opportunity to test the efficacy of such a device, before considering whether to incorporate it in future Mars missions.

Astronauts working on the Moon will be exposed constantly to a high-hazard environment. Clearly, for astronauts to remain physically healthy and psychologically stable it is paramount that crew health and safety be considered core issues. However, if health care measures fall short of ensuring the health and wellbeing of an astronaut and fail to prevent illness and/or injury, then medical intervention strategies must be implemented.

MEDICAL INTERVENTION

In February 2008 an undisclosed medical issue among the crew of the Space Shuttle Atlantis prompted a 24-hour delay to an EVA. ESA astronaut Hans Schlegel was eventually replaced by NASA astronaut Stanley Love and later rejoined the EVA rotation.

The incident was typical of the many minor medical conditions astronauts suffer during spaceflight. To date the spectrum of medical conditions (Table 6.5) reported by NASA and ESA astronauts have rarely required serious medical attention and there has been no medical evacuation of any NASA or ESA crewmember.

However, given the extreme nature of the lunar environment combined with the extended duration of a typical mission, it is inevitable that sooner or later medical intervention will be required to deal with one of more of the illnesses or injuries listed in Table 6.6.

Crew medical training

Faced with the possibility of dealing with either a Class II or Class III medical contingency, it is likely crewmembers will perform most necessary operations autonomously, with limited support from the ground. The reason for the requirement of autonomous health care while on the lunar surface is due to communication latency, which limits operational capabilities. While there is a one-second delay in communications to the ISS, communicating with astronauts on the Moon will result in a delay of several seconds, a situation compounded by possible intermittent loss-of-signal issues. Given these problems, not only must lunar-based health care maintenance be highly autonomous, it must also include "fail-safe" protocols and consist of resources dedicated to onsite medical care. This latter factor becomes increasingly important if the crew does not include a physician. Since it is not certain every mission will have a physician-astronaut, the burden of any in-mission medical contingency will fall upon the shoulders of the crew medical officer (CMO). At present the CMO is a pilot or scientist with 34 hours of medical training, whereas other crewmembers receive only 17 hours of preflight medical training. However, given the extended missions to the Moon, crew medical training may be increased and astronauts selected for Moon missions will follow a schedule similar to the one outlined in Table 6.7.

Table 6.5. Medical problems on-orbit.

Date	Nature of problem	Details
1969–1972	Minor infections	During the Apollo Program, 13 incidences of minor infections including stomatitis and pharyngitis, as well as recurrent inguinal and axillary infections were noted
1970	Urinary tract infection	After enduring freezing temperatures caused by losing power to the Command Module, Apollo 13 astronaut Fred Haise developed *Pseudomonas aeruginosa* urinary tract infection [3]
1971	Rapid decompression	During Soyuz 11's return from Salyut 1, a pressure equalization valve was jerked loose, depressurizing the cabin; cosmonauts Dobrovolskiy, Volkov, and Patsayev died
Skylab	Ventricular tachycardia	During a lower-body negative pressure protocol, one crewmember had a five-beat run of ventricular tachycardia [6]
1985	Fever	Cosmonaut Vladimir Vasyutin while onboard Salyut 7 became sick, and when his condition did not improve, returned to Earth early
1986	Dysrhythmias	Cosmonaut Alexander Laviekin, having spent six months on Mir, cut his mission short due to dysrhythmias
1990	Motion sickness	NASA astronaut James Bagian administered intramuscular injection to a crewmember for a disabling case of motion sickness [9]
1991	Cut	During Space Shuttle mission STS-37, a palm restraint on a crewmember's glove became loose and punched a hole in the pressure bladder between his thumb and forefinger; the astronaut bled out into space but the coagulating blood sealed the opening
1996–1997	Depression	NASA astronaut John Blaha experienced fits of anger, insomnia, and withdrawal caused by an excessive workload; with a reduced workload and support he completed his mission.

To ensure adequate treatment and rehabilitation during extended lunar missions, NASA will rely on instructing the crew using curricula and algorithms (Figure 6.5) based on microgravity physiological models of a human patient simulator (HPS). The curricula will include medical training as well as telementoring and telemedicine

Table 6.6. Classification of illnesses and injuries in spaceflight [13].

Characteristics	Examples	Type of response
Class I • Mild symptoms • Minimum effect upon performance • Non life-threatening	• Space motion sickness • Gastrointestinal distress • Urinary tract infection • Upper respiratory infection • Sinusitis	• Self-care • Administration of prescription and/or non-prescription medication
Class II • Moderate to pronounced symptoms • Significant effect upon performance • Potentially life-threatening	• Decompression sickness • Air embolism • Cardiac arrhythmia • Toxic substance exposure • Open/closed chest injury • Fracture • Laceration	• Immediate inflight diagnosis and treatment • Possible evacuation • Possible mission termination
Class III • Immediate severe symptoms • Incapacitating • Unsurvivable if definitive care unavailable	• Explosive decompression • Overwhelming infection • Massive crush injury • Open brain injury • Severe radiation exposure	• Immediate evacuation following resuscitation and stabilization if necessary • Comfort measures applied

techniques based on the high-fidelity environment analog training (HEAT) concept that recreates a patient care facility on the Moon.

Telemedicine

In the event of a serious injury requiring surgical intervention, it is probable that remote or telepresence surgery (i.e., telemedicine) will be required (Figure 6.6, see color section). By this method of surgery, a surgeon in a remote location controls the robotic instruments performing the actual surgery. This method, which has been practiced successfully over intercontinental distances, has also been investigated by NASA, conducting research aboard its undersea research station *Aquarius*, 20 meters below the ocean, near the coast of Key Largo, Florida.

"We have learned that it is possible, and quite safe, to telementor an untrained person through a complex medical task."

Dr. Mehran Anvari, Principal Investigator,
McMaster University Center for Minimal Invasive Surgery

Table 6.7. NASA medical training for International Space Station crewmembers [14].

Training session	Crew	Time	Time prior to launch
ISS space medicine overview	Entire crew	0.5 h	18 mo
Crew Health Care System (CHeCS) overview	Entire crew	2 h	18 mo
Cross-cultural factors	Entire crew	3 h	18 mo
Psychologic support familiarization	Entire crew	1 h	18 mo
Countermeasures System Operations 1	Entire crew	2 h	12 mo
Countermeasures System Operations 2	Entire crew	2 h	12 mo
Toxicology overview	Entire crew	2 h	12 mo
Environmental health system microbiology operations and interpretation	ECLSS	2 h	12 mo
Environmental health system water quality operations	ECLSS	2 h	12 mo
Environmental health system toxicology operations	ECLSS	2 h	12 mo
Environmental health system radiation operations	ECLSS	1.5 h	12 mo
Carbon dioxide exposure training	Entire crew	1 h	12 mo
Psychologic factors	Entire crew	1 h	12 mo
Dental procedures	CMOs	1 h	8 mo
ISS Medical Diagnostics 1	CMOs	3 h	8 mo
ISS Medical Diagnostics 2	CMOs	2 h	8 mo
ISS Medical Therapeutics 1	CMOs	3 h	8 mo
ISS Medical Therapeutics 2	CMOs	3 h	6 mo
Advanced Cardiac Life Support (ACLS) equipment	CMOs	3 h	6 mo
ACLS pharmacology	CMOs	3 h	4 mo
ACLS Protocols 1	CMOs	2 h	4 mo
ACLS Protocols 2	CMOs	2 h	4 mo
Cardiopulmonary resuscitation	Entire crew	2 h	4 mo
Psychiatric issues	Entire crew	2 h	4 mo

Training session	Crew	Time	Time prior to launch
Countermeasures System evaluation operations	CMOs	3 h	4 mo
Neurocognitive assessment software	Entire crew	1 h	4 mo
Countermeasures System maintenance	Entire crew	2.5 h	4 mo
Environmental Health System preventive and corrective maintenance	Entire crew	1 h	4 mo
ACLS "megacode" practical exercise	Entire crew	3 h	3 mo
Psychologic Factors 2	Entire crew	2 h	1 mo
Medical refresher	Entire crew	1 h	2 wk
CMO computer-based training	CMOs	1 h/mo	During mission
CHeCS contingency drill	Entire crew	1 h	During mission

During the NASA Extreme Environment Mission Operations (NEEMO) mission in October 2004, Dr. Anvari remotely guided the crew through gall bladder surgery and suturing of arteries while remaining in his hometown of Hamilton, Ontario. Although three of the six-member crew were physicians, none were surgeons. In addition to the assistance provided by Dr. Anvari, NEEMO's crew were aided by the Zeus System, a robot designed for the purpose of telemedicine.

The 2004 NEEMO mission and subsequent NEEMO missions have demonstrated telemedicine technology to be a tool capable of providing downlink of video and diagnostic procedural images (Figure 6.7, see color section) in addition to connecting a crewmember with qualified surgeons. However, despite the sophistication of telemedicine, technology will never be a substitute for the skills of a well-trained CMO and a motivated crew. Although it is possible to anticipate and prepare for common surgical and medical emergencies, unanticipated events do and can occur. For example, a subarachnoid hemorrhage was managed by the ingenuity and innovation of physicians in Antarctica, despite inadequate facilities [20]. Such an event not only underlines the human drive to save life against all odds, but also demonstrates how improvisation can lead to a successful outcome.

Throughout its history, NASA has always dealt successfully with transition. First, the transition from atmospheric flight to supersonic flight, then from suborbital to orbital flight, and from orbital flight to missions beyond Earth orbit. Long-duration missions beyond Earth orbit represent another transition fraught with formidable challenges for biomedical care. However, NASA has taken the steps to

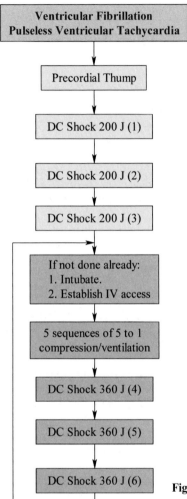

Figure 6.5. Diagnostic algorithm used to identify medical problems. Adapted from NASA.

ensure the transition to extended missions on the Moon is a successful one. The agency has learned from the experience of crewmembers working in analog environments such as Antarctic research stations and submarine missions and, perhaps most importantly, it has learned that the human factor will become ever more important as the durations of missions increase.

REFERENCES

[1] Ark S.V.; Curtis K. *Spaceflight and Psychology: Psychological Support for Space Station Missions.* Behavioral Health & Performance Group, NASA Johnson Space Center, Houston, TX (1999).

[2] Atkinson M.A.; Maclaran N.K. The pathogenesis of insulin dependent diabetes. *New England Journal of Medicine*, **331**, 1428–1436 (1994).

[3] Berry, C.A. *Biomedical Results of Apollo*, NASA-SP-368, p. 68. Scientific & Technical Information Office, Washington, D.C. (1975).

[4] Burrough, B. *Dragonfly*. Harper Collins, New York (1998).

[5] Cater, J.P.; Huffman, S.D. Use of Remote Access Virtual Environment Network (RAVEN) for coordinated IVA–EVA astronaut training and evaluation. *Presence: Teleoperators and Virtual Environments*, **4**(2), 103–109 (Spring 1995).

[6] Charles, J.B.; Bungo, M.W.; Fortner, G.W. Cardiopulmonary function. In: A.E. Nicogossian, C.L. Huntoon, and S.L. Pool (eds.), *Space Physiology and Medicine*, p. 268. Lea & Febiger, Malvern, PA (1994).

[7] Chung, J.; Harris, M.; Brooks, F.; Kelly, M.T.; Hughes, J.W.; Ouh-young, M.; Cheung, C., Holloway, R.L.; Pique, M. *Conference Proceedings: Exploring Virtual Worlds with Head-mounted Displays, Non-holographic Three-dimensional Display Technologies, Los Angeles, January 15–20, 1989.*

[8] Comet, B. Study on the survivability and adaptation of humans to long-duration interplanetary and planetary environments (HUMEX). *European Initiatives in Advanced Life Support Developments for Humans in Interplanetary and Planetary Environments*, ESA-TN-003. ESA, Noordwijk, The Netherlands (2001).

[9] Davis, J.R.; Jennings, R.T.; Beck, B.G. Comparison of treatment strategies for space motion sickness. *Acta Astronautica*, **29**(8), 587–591 (August 1993).

[10] Donovan, D.J.; Huynh, T.V.; Purdon, E.B.; Johnson, R.E.; Sniezek, J.C. Osteoradionecrosis of the cervical spine resulting from radiography for primary head and neck malignancies: Operative and nonoperative management (Case report). *J. Neurosurg Spine*, **3**, 159–164 (2005).

[11] Durante, M. Biological effects of cosmic radiation in low-Earth orbit. *International Journal of Modern Physics*, 125–132 (2002).

[12] Facorro, G.; Sarrasague, M.M.; Torti, H.; Hager, A.; Avalos, J.S.; Foncuberta, M.; and Kusminsky, G. Oxidative study of patients with total body irradiation: Effects of amifostine treatment. *Bone Marrow Transplant*, **33**, 793–798 (2004).

[13] Houtchens, B.A. Medical-care systems for long-duration space missions. *Clin. Chem.*, **39**(1), 13–21 (1993).

[14] Jennings, R.T.; Sawin, C.F.; Barratt, M.R. Space operations. In: R.L. DeHart, J.R. Davis (eds.), *Fundamentals of Aerospace Medicine*, Third Edition. Lippincott, Williams & Wilkins (2002).

[15] Kanas, N.; Manzey, D. *Space Psychology and Psychiatry*. Microcosm Press, El Segundo, CA/Kluwer Academic, Dordrecht, The Netherlands (2003).

[16] LeBlanc, A.; Schneider, V.; Shackelford, L.; West, S.; Ogavov, V.; Bakulin, A.; Veronin, L. Bone mineral and lean tissue loss after long duration spaceflight. *Journal of Bone Mineral Research*, **11**, S323 (1996).

[17] McCormick, T.J.; Lyons, T.J. Medical causes of inflight incapacitation: USAF experience 1978–1987. *Aviation Space and Environmental Medicine*, **62**, 882–887 (1991).

[18] Olshansky, B. Syncope: Overview and approach to management. In: B.P. Grubb and B. Olshansky (eds.), *Syncope: Mechanisms and Management*, p. 18. Futura, Armonk, NY (1998).

[19] Palinkas L.A. Group adaptation and individual adjustment in Antarctica: A summary of recent research. In: A.A. Harrison, Y.A. Clearwater, and C.P. McKay (eds.), *From Antarctica to Outer Space: Life in Isolation and Confinement*, pp. 239-251. Springer-Verlag, New York (1991).

[20] Pardoe, R.A Ruptured intracranial aneurysm in Antarctica. *Medical Journal of Australia*, **1**, 344–350 (1965).

[21] Rebo, R.K.; Amburn, P. A helmet-mounted environment display system. In: *Helmet-Mounted Displays*. SPIE, Bellingham, WA [*Proc. SPIE*, **1116**, 80–84 (1989)].

[22] Sandal, G.M.; Vaernes, R.; Bergan, T.; Warncke, M.; Ursin, H. Psychological reactions during polar expeditions and isolation in hyperbaric chambers. *Aviation, Space and Environmental Medicine*, **67**, 227–234 (1996).

[23] Santy, P.A. *Choosing the Right Stuff. The Psychological Selection of Astronauts and Cosmonauts*. Praeger, Westport, CT (1994).

[24] Santy, P.A.; Holland, A.W.; Faulk, D.M. Psychiatric diagnoses in a group of astronaut candidates. *Aviation, Space and Environmental Medicine*, **62**, 969–973 (1991).

[25] Slater, M.; Steed, A.; McCarthy, J.; Maringelli, F. The influence of body movement on subjective presence in virtual environments. *Human Factors*, **40**(3), 469–477 (1998).

[26] Stuster, J.C. *Bold Endeavours: Lessons from Polar and Space Exploration*. Naval Institute Press, Annapolis, MD (1996).

[27] Vogel, J.M.; Whittle, M.W. Bone mineral content changes in the Skylab astronauts. *Proc. Am. Soc. Roentgenol.*, **126**, 1296–1297 (1976).

[28] Weybrew, B.B.; Noddin, E.M. Psychiatric aspects of adaptation to long submarine missions. *Aviation, Space and Environmental Medicine*, **50**, 575–580 (1979).

[29] Wolf, P.A.; Cobb, J.L.; D'Agostino, R.B. Pathophysiology of stroke: Epidemiology of stroke. In: H.J.M. Barnett, J.P. Mohr, B.M. Stein, and F.M. Yatsu (eds.), *Stroke: Pathophysiology, Diagnosis and Management*, Second Edition, p. 4. Churchill Livingstone, New York (1992).

7

From launch to landing

*We can continue to try and clean up the gutters all over the world and spend all of
our resources looking at just the dirty spots and trying to make them clean.
Or we can lift our eyes up and look into the skies and move
forward in an evolutionary way.*

Buzz Aldrin

MISSION ARCHITECTURE

The phrase *mission architecture* is one often used by engineers in deciding how to
piece together the elements required for a space mission, whether that mission
involves sending robotic explorers to Mars or launching astronauts into low Earth
orbit (LEO). Mission architecture also describes a combination of modes specific to
the mission, the assigning of flight elements such as the Launch Vehicle (LV) and, in
the case of establishing a lunar outpost, a description of the activities performed on
the lunar surface. What follows is a description of the process resulting in the choice
of the mission mode that will return humans to the Moon.

Lunar mission modes

The lunar mission mode (LMM) is an essential lunar architecture decision defining
how spaceflight components are integrated and the functions of each of these components. LMM constitutes an analysis of the mission-specific technology through a
process of refinement and concept exploration directed at validating a particular
mission mode, hence the name. To determine LMM, NASA conducts studies to
better understand the mission requirements for human exploration of the Moon in

the context of research and development (R&D) programs. By conducting these types of studies, NASA is able to define a baseline against which other mission-specific concepts are evaluated. For example, the results from these studies help NASA to define mission components such as flight experiments, as well as enabling the agency to communicate the feasibility of the LMM to the media, the public, and stakeholders.

EVOLUTION OF THE CURRENT LUNAR ARCHITECTURE

In preparation to conduct the LMM for the return to the Moon, NASA scrutinized similar studies characterizing the configuration of the Apollo capsule and the architecture employed by the Apollo missions. NASA also reviewed more recent lunar mission studies, such as the Office of Exploration (OExP) *Lunar Evolution* study of 1989 [8], the *Human Lunar Return Study* of 1996 [6], and the *America at the Threshold* study of 1991, which included chapters titled "Moon to stay", and "Mars exploration" [1]. Since many of the recommendations of these studies are incorporated into the current LMM, it is useful to review some of the more significant ones.

Space Exploration Initiative

"First, for the coming decade, for the 1990s, Space Station Freedom, our critical next step in all our space endeavors. And next, for the next century, back to the Moon, back to the future, and this time to stay. And then a journey into tomorrow, a journey to another planet, a manned mission to Mars ..."

President George Bush, July 21st, 1989

With these words, President George Bush launched the Space Exploration Initiative (SEI) and a 90-day study outlining mission architectures for a return to the Moon [2]. In the 1989 version of returning to the Moon it was envisaged the Space Shuttle would transport astronauts and cargo to Space Station Freedom (later to become the International Space Station) where they would transfer to a Lunar Transfer Vehicle/ Lunar Excursion Vehicle (LTV/LEV). The LTV/LEV would have been delivered to Freedom by Shuttle-derived (Shuttle C) unmanned launch vehicles. For each lunar landing, three Shuttle C launches would be required of which one would be for the spacecraft and two for the fuel. The architecture required two Shuttle C variants, one the standard version designed to deliver propellant, and another version designed to accommodate a 7.6-meter diameter payload shroud to launch the LEV.

For the trip between Freedom and lunar orbit, a crew of four astronauts would use the LTV, based on a design by Boeing following a Phase A contract in June 1989. Propelling the LTV would be four expendable oxygen and hydrogen fuel tanks, designed to be discarded in Earth and lunar orbit. The second element of the lunar transportation system was the LEV, which would transport crews and cargo from lunar orbit and the lunar surface.

The SEI architecture for establishing the lunar outpost required two unmanned test flights, one to deliver an unpressurized rover and equipment to prepare the outpost site, and a second to deliver a permanent habitat derived from Freedom's habitation module. If all had gone to plan, it was anticipated the first manned mission was to take place in 2001. A second crew was scheduled to arrive in early 2002 for the first six-month increment and in mid-2002 a third crew, which would stay for a full 12 months.

Office for Exploration

Other architecture studies conducted by NASA were conducted under the auspices of the agency's Office for Exploration (OExP), established in June 1987 in response to the 1986 National Commission on Space task force report led by astronaut Sally Ride. The task force report *Leadership and America's Future in Space* [7] outlined four initiatives, one of which included human exploration of the solar system. Subsequent to the publication of the task force report, the OExP [8] conducted a number of studies directed at defining human exploration beyond LEO, one of which was titled *Lunar Outpost to Early Mars Evolution*. The first set of studies conducted in 1988 served an important function by establishing the boundaries of various mission-specific conditions, as well as refining future mission criteria. The studies also resulted in several recommendations that were to become important when it came to determining how to return humans to the Moon. For example, it was decided the International Space Station (ISS) was key to developing the capability to live and work in space for extended periods and heavy-lift transportation would need to be pursued, with a capability of transporting large quantities of mass to LEO.

The 1988 studies led the OExP to continue their investigation by conducting additional case studies the following year, one of which addressed the topic of lunar evolution, which covered such broad subject areas as life support systems and lunar liquid oxygen (LOX) production. The case studies provided NASA with a baseline for reference architectures, the details of which were incorporated into the Synthesis Group's *America at the Threshold* study in 1991 [1]. *America at the Threshold* investigated candidate architectures for the Moon and identified key supporting technologies for future exploration such as automated rendezvous and docking (AR&D), zero-*g* countermeasures, radiation effects and shielding, and lightweight structural materials.

First Lunar Outpost study

The Synthesis Group's study was followed two years later by the First Lunar Outpost (FLO) study [5] which sought to identify the technical, programmatic, and budgetary implications of restoring NASA's lunar exploration capability. The outcome of the FLO study was to identify certain key features that were ultimately to become a part of the lunar architecture 15 years later. For example, the FLO study adopted an evolutionary approach with regard to minimizing operational complexity and suggested the most viable mission strategy involved a direct ascent to the lunar surface and a direct return to Earth.

Human Lunar Return study

In 1995, the *Human Lunar Return (HLR) Study* [6] was conducted at Johnson Space Center (JSC), its mandate being to reduce the cost of a human lunar mission by one to two orders of magnitude below previous human exploration estimates. Although the HLR represented a minimum mission approach for a return to the Moon, the study did propose a key element that was ultimately integrated into the current VSE lunar architecture, namely using the ISS as a staging node.

Decadel Planning Team

The next phase of planning was conducted under the auspices of an internal NASA task force designated the Decadel Planning Team (DPT), created with the purpose of defining an integrated vision and strategy for space exploration. The DPT eventually evolved into the NASA Exploration Team (NExT) and was responsible for identifying roadmaps for topics ranging from space transportation to revolutionary concepts such as electromagnetic launchers and nanostructures, although many of the more cutting edge concepts were ultimately proven to be unfeasible for human exploration missions. Although the focus of the NExT group was more revolutionary than evolutionary, the group analyzed and refined mission architectures conducted by previous groups. Among their conclusions was that lunar missions utilizing existing launch capabilities were feasible and would serve a useful function in terms of defining technologies and operational concepts. Perhaps most significantly, NExT laid the foundation and framework to develop NASA's Exploration Blueprint Team and, ultimately, NASA's Integrated Space Plan (ISP), which collectively embraced the "stepping stone" approach to exploration beyond LEO.

Exploration Systems Mission Directorate

Although the momentum of the Exploration Blueprint Team and the ISP was slowed by the Columbia tragedy in February 2003, the requirements made as a result of the studies conducted were implemented under the auspices of NASA's Exploration Systems Mission Directorate (ESMD). ESMD was created in January 2004 [4], in response to President Bush's new Vision for Space Exploration (VSE). Incorporated within the VSE were lunar missions designed specifically to support longer term exploration such as manned missions to Mars and beyond. Specifically, the ESMD focused upon key components of the exploration architecture, namely lunar landing sites, mission modes, propulsion, Earth landing modes and heavy-lift concepts such as Evolved Expendable Launch Vehicles (EELVs). It was at this juncture that the lunar architecture began to take shape, as ESMD awarded a series of contracts with the goal of defining vehicle concepts and mission architecture designs. Based on the outcome of the contracts the currently accepted VSE lunar architecture was developed. The following section provides a step-by-step description together with the rationale for each component.

LUNAR ARCHITECTURE

Robotic precursor missions

The first step in returning humans to the Moon is to send robotic probes to research the hostile environment and to identify potential landing sites.

Lunar Reconnaissance Orbiter

The first of the robotic probes, scheduled for launch on an Atlas V 401 rocket in late 2008, is the Lunar Reconnaissance Orbiter (LRO). The LRO's task will be to create an atlas of the Moon's features and resources. After a four-day transit from the Earth to the Moon, the LRO will first enter a commissioning orbit and then a polar orbit which will take it as close as 50 km above the lunar surface. The LRO will carry a comprehensive payload (Table 7.1) enabling it to provide information essential to a safe human return to the Moon.

Co-manifested aboard the Launch Vehicle for the LRO will be the Lunar Crater Observation and Sensing Satellite (LCROSS), tasked with detecting the presence or absence of water ice in the permanently shadowed crater at the Moon's South Pole. LCROSS and the LRO are scheduled to launch aboard an Atlas V rocket at the end of 2008. After launch the Atlas V's Centaur upper stage will perform a fly-by of the Moon and enter an elongated Earth orbit to position LCROSS for impact on the South Pole. Approaching the Moon, the spacecraft and upper stage will separate and Centaur will impact the lunar surface, creating a debris plume which will be analyzed by specialized instruments aboard the LCROSS.

Together, the LRO and LCROSS spacecraft will provide crucial data for lunar mapping and modeling enabling the establishment of a lunar outpost in preparation for human exploration.

Earth-to-orbit transportation

Following the precursor missions, the next step will be to send humans to the Moon. Studies conducted by NASA over the years have examined a variety of Launch Vehicles (LVs) including existing capabilities such as EELVs and yet-to-be-designed Heavy Lift Launch Vehicles (HLLVs). Key considerations for mission planners when deciding on the type of Earth-to-orbit (ETO) transportation mode included the mass to be lifted into LEO, on-orbit assembly requirements, and LV shroud volume. Based on the studies to date, mission planners decided using an EELV was only marginally feasible, very complex, and therefore operationally challenging to the point where it was ultimately discounted. Since the initial mass in LEO required for a manned lunar mission was calculated to be between 120 t and 220 t, it was determined the payload

Table 7.1. Lunar Reconnaissance Orbiter instrument payload.

Instrument	Acronym	Function
Cosmic Ray Telescope for the Effects of Radiation	CRaTER	• Characterize lunar radiation environment • Measure radiation absorption
Diviner Lunar Radiometer Experiment	DLRE	• Measure surface and subsurface temperatures • Identify cold traps and potential ice deposits
Lyman Alpha Mapping Project	LAMP	• Map lunar surface in ultraviolet spectrum • Search for surface ice • Provide images of shadowed regions
Lunar Exploration Neutron Detector	LEND	• Create high-resolution maps of hydrogen distribution • Characterize neutron component of lunar radiation environment
Lunar Orbiter Laser Altimeter	LOLA	• Measure landing site slopes • Create three-dimensional map of the Moon
Lunar Reconnaissance Orbiter Camera	LROC	• Take black-and-white images of surface • Take color and ultraviolet images of lunar surface at 100 m resolution
Mini-RF Technology Demonstration	Mini-RF	• Image the polar regions to search for water ice • Demonstrate ability to communicate with an Earth-based ground station

mass and volume could be met with concepts derived and evolved from existing Space Shuttle systems. The choice of such a system meant mission planners were limited in their choice of mission modes due to the payload limitations. Had a single LV been in existence with the capability of inserting 200 metric tonnes into LEO, mission planners would have been able to consider a *direct–direct* architecture in which the vehicle would be launched directly from the Earth to the Moon, with no Earth orbit rendezvous (EOR) and no lunar orbit rendezvous (LOR). Instead, because of the cost-influenced choice of a system evolved from the Shuttle mission, planners were left with three mission modes to choose from, each of which is evaluated in the following section.

Mission mode evaluation, Analysis Cycle 1

Mission mode evaluation was conducted in three stages by a team of mission planners. Each stage compared parameters such as cost, reliability, and safety and incorporated an index to compare mission options. Following each stage, certain mission modes were eliminated or subjected to further studies to evaluate why a particular mission mode should be accepted or rejected. The baseline reference mission subject to evaluation was the LOR split mission architecture, also termed the ESAS Initial Reference Architecture (EIRA), described in detail in the following sections [3].

The first evaluation of the EIRA was a comparison of Earth orbit rendezvous–lunar orbit rendezvous (EOR–LOR), EOR–direct return, and EOR–LOR variant, requiring the CEV to land on the lunar surface. The team, which represented the analytical core of NASA, evaluated the mission modes by considering major mission factors such a flight mechanics, reliability, safety, and cost. After each factor had been considered and assessed, the team generated a mission performance analysis including anticipated integrated program costs, and calculated the probability of losing a crew (P/LOC) and the probability of losing a mission (P/LOM).

Once the analysis had been performed, the EIRA team defined a sortie mission (Figure 7.1), returning four astronauts to the Moon in 2018 for a seven-day stay. The initial seven-day mission would then be followed by missions every six months, until the lunar outpost was established sometime in 2022.

The EIRA team decided functional elements of the outpost (such as power systems, resource utilization equipment, and habitat module) would be delivered by dedicated cargo landers in a series of flights, commencing in 2020 and continuing

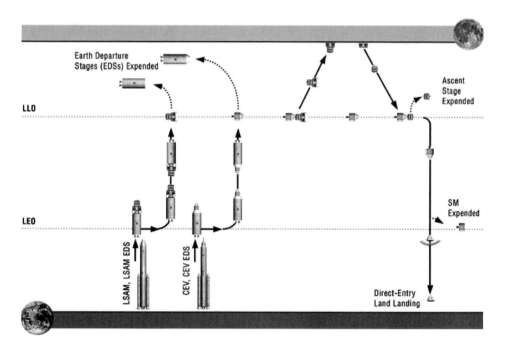

Figure 7.1. ESAS Initial Reference Architecture. Image courtesy: NASA.

to the end of 2021. Finally, it was envisaged the lunar outpost would achieve operational capability in 2022, by which time the first outpost crew would arrive ready to commence a 6-month rotation on the lunar surface. As with current ISS rotations, a new crew would arrive every six months for the duration of the outpost's lifetime, which was anticipated to remain operational until at least 2030.

Other considerations the EIRA team addressed during the first analysis cycle included configuration studies to determine flight elements and launch masses for the different mission modes in an attempt to decide on the most effective mode. The configuration studies required a complex understanding of the crew-carrying elements, such as the CEV and the Lunar Lander, as well as determining the minimum habitable volume in the CEV with and without the Shuttle Advanced Crew Escape Suit (ACES). Many of the configuration studies were conducted by active and non-active astronauts, each of whom had experience either onboard the Space Shuttle, ISS, or Mir. To assist the astronauts in their configuration studies, a low-fidelity mock-up of the CEV was constructed at JSC (Figure 7.2, see color section). The 14.9 m^3 volume of the mock-up was then fitted with systems and seats to allow the astronauts to gauge if there was sufficient space for the zero-g activities that would be required.

Other aspects of the first analysis cycle included an assessment of factors such as vehicle mass properties, safety, reliability, and likely mission-critical events, to determine which of the mission modes would be the most effective. For example, the EIRA team considered a mission requiring vehicle rendezvous and docking only while in low lunar orbit (LLO), whereas another option assessed required vehicle rendezvous and docking only in Earth orbit. After several comparisons of mission modes and their variants, the EIRA team identified four primary mission architectures for further evaluation, these being LOR, EOR–LOR, EOR–direct return, and EOR–LOR with CEV to surface (also known as EOR–variant), each of which are described in the following section.

LOR architecture

LOR (Figure 7.3) is an architecture in which the LSAM is pre-deployed in a single launch to LLO. The crew and CEV are then launched to lunar orbit using the same vehicle type whereupon the two vehicles rendezvous and dock. The crew then transfers to the LSAM, which undocks from the CEV and descends to the surface, while the CEV, CM, and SM remain unmanned in LLO.

After their seven-day surface stay, the LSAM returns the crew to LLO and docks with the CEV. The crew then transfer to the CEV, and the CEV returns to Earth.

In the Analysis Cycle 1 study, the proposed CEV is designated as a four-astronaut capsule with a pressurized volume of 22.4 m^3 providing 47 crew-days of life support capability and includes high-density polyethylene (HPDE) shielding on the sidewalls and ceiling for radiation protection. The CEV SM is designated as an unpressurized cylinder designed to contain the propulsion and power generation systems for the vehicle. To enable orbital maneuvering such as rendezvous and

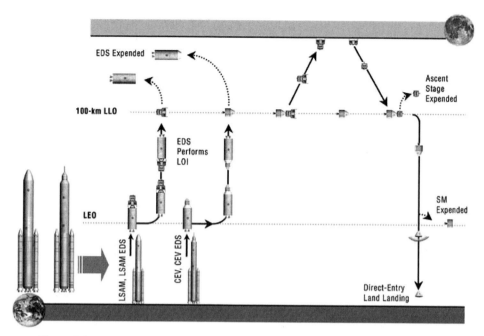

Figure 7.3. Lunar orbit rendezvous mission architecture. Image courtesy: NASA.

docking, the SM is fitted with two pressure-fed engines and twenty-four reaction control thrusters (RCTs). All told, the combined LEO mass of the CM and SM is 22,909 kg.

The configuration for the Analysis Cycle 1 LSAM is based on the requirement to transport four astronauts from LLO to the lunar surface and to support the crew for up to seven days, before returning them to the CEV in LLO. In a design echoing the Apollo LM, the LSAM includes separate ascent and descent stages and is fitted with the capability to fully depressurize the ascent stage crew cabin to permit extra-vehicular activity (EVA). To enable ascent from the lunar surface and to perform orbital maneuvering, the LSAM is fitted with two 22.2 kN pressure-fed main engines and sixteen 445 N RCTs providing a capability to perform a 1,882 m/s delta-V. The combined LSAM mass including the ascent stage is 27,908 kg, of which 18,010 kg comprises the descent stage. The latter stage also incorporates 500 kg of payload such as science equipment destined for the lunar surface.

EOR–LOR architecture

In this architecture, basically a variant of the LOR architecture, the CEV–LSAM docking occurs in LEO rather than LLO and, although the architecture still incorporates two launches, instead of delivering the CEV and LSAM to the Moon in two launches, one launch is designated for a large EDS and one launch for the CEV, crew, and LSAM. Another difference between EIRA and EOR–LOR is that the latter

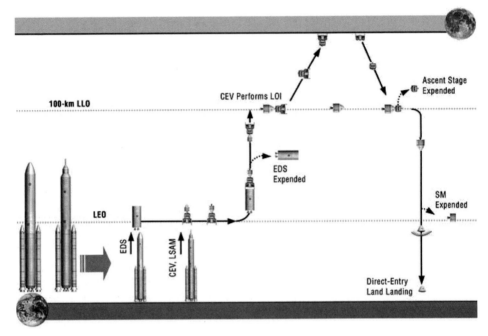

Figure 7.4. Earth orbit rendezvous–lunar orbit rendezvous mission architecture. Image courtesy: NASA.

requires the combined CEV and LSAM to dock with the EDS in Earth orbit, where-upon the EDS performs TLI. Apart from these differences, the architecture is the same as EIRA (Figure 7.4).

In terms of the actual flight components there is little change except that the EOR–LOR CEV permits 53 crew-days of life support capability, which is a function of the additional orbital maneuvering demanded by this architecture. Another difference is the design of the SM, which is required to carry a higher propellant quantity than in the EIRA architecture because the LSAM is attached to the CEV during the LOI maneuver. The major difference between the EIRA and EOR–LOR architectures, however, is the difference in total component mass since the latter total combined CEV mass is 59,445 kg vs. a combined CEV mass of 22,909 kg for the EIRA configuration.

EOR–direct return architecture

This architecture is very different from the previous two since, unlike the LOR architecture, it requires no rendezvous maneuvers once the CEV and LSAM depart LEO. Another distinguishing feature of the LOR architecture is leaving some part of the CEV in LLO, while a lunar-landing system transports the astronauts between LLO and the surface of the Moon. On the other hand, the EOR–direct return architecture carries the whole Earth return configuration all the way to the surface

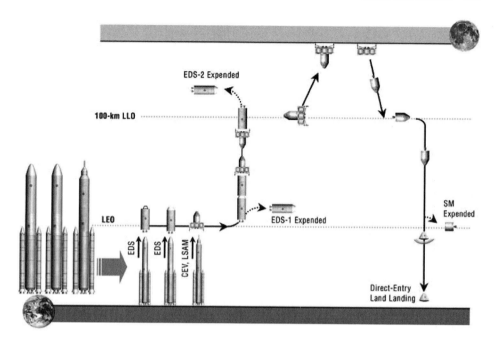

Figure 7.5. Earth orbit rendezvous–direct return mission architecture. Image courtesy: NASA.

of the Moon. This architecture significantly simplifies the mission but, despite the greater elegance of this mission configuration, this simplicity is not achieved without cost since it requires a third launch. One argument for a third launch and the concomitant increase in cost and LEO architecture is made for the inherent flexibility afforded by the EOR–direct return configuration where the astronauts spend the entire mission in the CEV, thereby canceling out the requirement for two crew cabins. As can be seen in Figure 7.5, the mission mode for the EOR–direct return architecture requires two EDSs to deliver the vehicles to LLO. The EDSs comprise the first two launches of the mission and are automatically docked in LEO prior to the crew, CEV, and LSAM being launched in the third launch. The vehicles then dock to the EDSs and the TLI is performed. Before TLI is completed the first EDS is depleted and subsequently disposed of and the second EDS completes TLI, followed four days later by LOI. Following LOI, the CEV and LSAM land on the Moon and, after a seven-day stay on the lunar surface, the CEV returns the crew directly to Earth.

Since the CEV in the EOR–direct return architecture was required to provide habitable volume for a seven-day lunar surface stay, the vehicle was modified to provide $39.0\,m^3$ of pressurized volume in addition to the supplementary displays and controls to enable astronauts to land and take off from the Moon's surface. Also added were full-cabin depressurization and extra life support capabilities to permit multiple EVAs. Another modification was the configuration of three 44.5 kN pressure-fed main engines and sixteen 445 N RCTs providing 2,874 m/s ascent, TEI, and delta-V.

The role of the LSAM in the EOR–direct return architecture is simply to perform a powered descent for the astronauts and CEV from LLO to the surface of the Moon, which it executes using a propulsion system comprising five 44.5 kN pressure-fed main engines and sixteen 445 N RCTs. Such a configuration enables the LSAM to perform 2,042 m/s of LEO rendezvous and attitude control delta-V.

EOR–LOR with CEV-to-surface architecture

The fourth architecture considered constitutes a hybridized version of the EOR–direct return and the EOR–LOR architectures, utilizing the LOR component of the latter with the single crew volume of the former. In this architecture, instead of the CEV CM and SM remaining in LLO, only the SM remains, permitting the use of the CM for lunar surface operations. Unfortunately, as with the EOR–direct return architecture, the EOR–LOR with CEV-to-surface configuration incurs the penalty of a third-launch requirement since the combined CEV and LSAM mass exceed the capabilities of a single EDS.

Until the point of powered descent to the surface of the Moon, the EOR–LOR with CEV-to-surface architecture is identical to the mode utilized by the EOR–return format. At the point of powered descent, the architecture deviates since the CEV separates from the LSAM and the CM separates from the SM. The CM then returns to the LSAM and docks to the ascent stage, utilizing a docking module positioned beneath the aft heat shield of the CM. Once all these maneuvers have been completed, the LSAM transports the crew to the Moon's surface for a seven-day stay, the ascent stage then returns the crew and CEV CM to LLO, whereupon the vehicles separate and the CM docks with the SM. From this point in the architecture onwards the sequence of events is identical to those in the EOR–LOR architecture (Figure 7.6).

Except for the addition of an extra four days of crew life support capability, the only major difference between the EOR–LOR with CEV-to-surface and the EIRA architecture is the addition of a docking module to the CEV, enabling docking to the ascent stage and additional command and control (C&C) avionics in the SM. Few modifications are required in the configuration of the LSAM except for the difference in propellant loading, a consequence of the lower landed mass of the descent stage.

Architectures compared

Upon completing Analysis Cycle 1, each architecture was assessed and compared by evaluating architecture criteria such as mass breakdowns for each subsystem, and a summary of mission events and mission-critical events such as orbital maneuvers and engine burn times. In addition to functionality, the team defined anticipated life cycle costs until the year 2025, which were compared with the EIRA baseline mission. Other important comparisons were made related to risk and reliability in terms of the likelihood of P/LOC or P/LOM. Once configuration considerations and other aspects of the architectures had been evaluated, the team assessed the advantages and disadvantages of each architecture. Certain aspects were common to all architectures such as the configuration of the CEV, varying in diameter from 5.0 m to 5.5 m, and the 1,400 kg of supplemental radiation protection. Another factor holding constant in

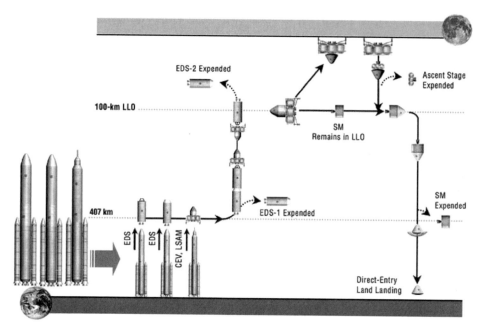

Figure 7.6. Earth orbit rendezvous–lunar orbit rendezvous with Crew Exploration Vehicle to surface mission architecture. Image courtesy: NASA.

the comparison of architectures was the propulsion type and technology utilized for LOI, lunar descent/ascent, and TEI.

Based on the aforementioned criteria, it was determined the architectures incurring the lowest cost and the fewest launches were LOR mission modes, whereas the direct return mission modes required the highest number of launches and also incurred the highest P/LOM and lowest P/LOC. Of all the architectures, the one performing the worst was the EOR–LOR with CEV-to-surface, incurring the highest cost and also having the highest P/LOM. These factors resulted in this architecture being eliminated from further evaluation.

Once all architectures had been evaluated the team moved on to Analysis Cycle 2 which was to focus upon means of optimizing mission mode performance and evaluating factors such as radiation shielding.

Mission mode evaluation, Analysis Cycle 2

Based on the outcome of Analysis Cycle 1, the goal in Analysis Cycle 2 was twofold. First, the team sought to optimize architecture performance by examining propulsion technology, CEV systems, and mass of radiation shielding. Second, due to the stress imposed upon the CEV in the EOR–direct return architecture, the team decided this particular mission mode would need to be subjected to closer scrutiny. The additional focus upon the EOR–direct return would determine if the CEV could function as a single crew compartment for landing, surface habitation, and ascent.

One of the first variables addressed by the team was the issue of mass savings, which inevitably resulted in the team assessing the CEV supplemental radiation shielding and propulsion technology in an attempt to identify where the greatest mass savings could be achieved. As a part of the assessment the ESAS team conducted a study evaluating the 1,800 kg supplemental radiation shielding

Radiation protection

Protecting the crew against radiation is one of the most critical aspects of any lunar architecture since, at a minimum, astronauts will spend nine days in the interplanetary radiation environment beyond the Earth's protective magnetosphere. During the translunar cruise, lunar orbit, and lunar surface stay phases of the mission, crews will be subject to ionizing radiation in the form of galactic cosmic rays (GCRs) and hydrogen and helium ions ejected by solar particle events (SPEs), all of which will inflict various direct and indirect injuries upon the astronauts in the form of either cell death or an increased predisposition for cancer. In addition, radiation in the form of worst-case SPEs may restrict surface EVAs and may, in certain cases, require an abort to orbit. Also, as lunar missions are extended beyond seven days, the issue of previous exposure becomes a factor in crew selection. Unsurprisingly, NASA has calculated both annual and lifetime limits for the amount of radiation an astronaut may be exposed to, although these limits apply only to missions in LEO. Since NASA has not conducted any interplanetary missions since Apollo, it does not have sufficient data to accurately predict physiological exposure effects during a mission to the Moon. Given the uncertainty in estimating radiation risk, NASA can only apply risk estimation and design vehicles that keep exposures "as low as reasonably achievable" (ALARA).

Obviously, future manned Moon missions will afford the ESMD Space Radiation Program an opportunity to collate data concerning radiation risks and how best to protect the health of astronauts. To that end, radiation research and protection programs have been devised which include objectives such as establishing a knowledge base on radiation limits, developing shielding tools, accurately estimating crew risks, and developing and integrating physiological countermeasure equipment.

For the purposes of the study, the ESAS team used NASA's established LEO limits. These limits were combined with risk projections and risk estimates based on the risk of exposure-induced death (REID) and the relationship between the anticipated shielding mass, dosage, and the probability of a significant radiation event, such as an SPE, occurring during a mission. Based on historical data, the team calculated three threshold levels of radiation risk, the first being an assessment of the probability an event would exceed current LEO limits within the mission timeframe, which was estimated at 0.2% [3]. The second threshold level was an estimation of the probability an event would debilitate and incapacitate the crew to the extent that 50% of the crew would vomit within 48 hours. This threshold was estimated to be 0.03% [3]. Finally, the team estimated the probability of a catastrophic event causing death within 30 days, which was estimated to be 0.01% [3].

Armed with this information, the team turned their attention to the design of the CEV and where best to place the components of the HDPE radiation shield to provide maximum protection for the astronauts. Using computer-assisted design (CAD) software and computer-aided manufacturing (CAM) models, the team estimated the radiation doses for the skin, eyes, and blood-forming organs (BFOs). Based on this calculation, the team, considering factors such as age, gender, shielding material, and exposure history, were able to estimate whether a crew would be safe inside the CEV with or without supplemental shielding. After crunching the numbers, the team discovered that without supplemental shielding the crew would not survive and, even with the addition of extra shielding, the LEO limit would exceed the upper 95% confidence interval [3]. This left engineers with a problem since the radiation shielding represented such a significant mass penalty in the architecture as it needed to be carried throughout the mission. Yet, even with the extra shielding the crew would not be adequately protected. So, unsurprisingly, further modeling was recommended as a part of Analysis Cycle 3.

Propellant analysis

The next architecture consideration was choice of propellant and propellant combinations for the CEV SM service propulsion system and the ascent propulsion system for the LSAM. Based on previous studies the candidate propellant combinations for the CEV SM included LO_2/LH_2, LO_2/LCH_4, LO_2/ethanol (EtOH), and monomethyl hydrazine (MMH)/nitrogen tetroxide (NTO), whereas for the LSAM the team had a choice between LOX/methane (CH_4) and MMH/NTO.

In choosing the best propellant combinations, the team had to consider a number of factors. These included previous flight history, mass penalties, whether the propellant ignited hypergolically (thereby eliminating the need for igniters), the toxicity of the propellants, and the degree to which the propellants caused an insidious process known as flow decay. The latter process, routinely observed on the Space Shuttle, is a consideration when choosing an NTO system since iron nitrate formation often occurs in such a system as moisture is absorbed by the propellant. Such an event results in nitric acid attacking and leeching iron from the iron alloy lines inside the engine. The result is the formation of iron nitrates which occlude flow passages. Inevitably, as the flow passages are occluded the propellant flow is reduced, resulting in an off-mixture ratio combustion and ultimately in non-ignition of the engine. However, the team decided the problems associated with flow decay were outweighed by the 40-year experience NASA has of operating NTO/MMH systems, and the propellant combination was therefore deemed to be a low-risk choice.

The LOX/CH_4 propellant assessed for use by the SM had no flight history and almost no history of ground testing. However, due to the clean-burning nature of the propellant, combined with its above-average combustion performance and the extensive use of LOX during on-orbit operations, the LOX/CH_4 was a serious candidate for consideration. Using CAD modeling, the team evaluated the various performance aspects of LOX/CH$_4$ but ultimately decided, due to the aggressive development schedule for the CEV, the development of a LOX/CH$_4$ system for the SM would not be feasible.

The team then turned to LOX/LCH$_4$, a propellant also without any flight history and only limited ground testing. However, due to the high specific impulse (I_{SP}) of LOX/LCH$_4$ combined with the light mass and low volume of the propellants propulsion system, LOX/LCH$_4$ was subjected to closer scrutiny. The team knew from limited ground testing that one of the problems inherent in any system using LOX/LCH$_4$ was restarting the engines following dormant periods and, because the temperature in LLO would drop to below $-100°$C, the team was forced to consider the use of heaters to preheat the propellants in order to achieve start conditions. Another problem faced by the team was liquid acquisition due to the propensity for LOX to "boil off" while in cryogenic storage. Given these problems, the team decided that before a propellant could be chosen, a prototype system would need to be constructed, not only to resolve the heating and boil-off problems but also to identify all risk areas.

Surface CEV studies

Perhaps the most elegant and therefore most attractive mission mode was the EOR–direct return due to the requirement for the CEV to descend to the lunar surface and function as the crew's habitat, a mission mode not requiring the development of a second crew cabin for the LSAM. In this architecture the CEV serves as a cabin for the astronauts during Earth ascent, LEO rendezvous and docking, translunar cruise, lunar orbit operations, trans-Earth cruise, and Earth entry. Although such a design affords the advantages of reduced risk and reliability due to fewer vehicles, it required the CEV to perform a number of unique functions yet to be evaluated. Also, because of these functions the EOR–direct return CEV would require a larger habitable volume than LOR-based CEVs due to the requirement of accommodating four astronauts on the surface of the Moon. To accurately define the amount of habitable volume required, NASA's Habitability and Human Factors Office (HHFO) conducted a series of CEV and LSAM volume studies using a layout of the CEV pressure vessel.

By studying these layouts, engineers were not only able to analyze volume and stowage requirements but were also able to investigate various design configurations such as positioning the airlock on the top and side of vehicle. In addition to the basic layout studies, the engineers also performed layout analyses of the CEV and LSAM during different phases of the mission such as lunar and lunar surface operations. This enabled them to define aspects pertaining to command, control, and telemetry interfaces between the CEV and LSAM and also to identify dust mitigation strategies.

The team also experimented with placing avionics equipment in different locations and evaluated the effect of eliminating hard-walled partitions in an effort to maximize the habitable volume of the vehicles. Finally, after evaluating and analyzing numerous permutations of equipment location and vehicle configurations, the engineers decided on the final layout described in Chapter 4.

Airlock

After resolving the configuration of the CEV and LSAM, the engineers turned their attention to the question of whether to include an airlock in the design of the LSAM. The Apollo design Lunar Lander was not fitted with an airlock and resulting lunar dust ingress posed a hazard not only to hardware but also to crew health, causing respiratory and eye irritation for many astronauts. Apart from preventing ingress of dust, an airlock would also protect the crew in the event of a suit malfunction or an injury, as well as permitting a more flexible work schedule, allowing split crew operations with one team conducting EVAs and the other remaining in the LSAM. In order to evaluate the advantages and disadvantages of the LSAM with and without an airlock, the engineers assessed all possibly occurring hazards and contingencies on the lunar surface and then designed solutions. The solutions were then accorded a risk level and based on the risk levels with or without an airlock, it was deemed advisable to include an airlock in the LSAM configuration due to the high risk of conducting lunar surface operations without one.

Next on the Analysis Cycle 2 agenda was an evaluation of the LVs used for the EOR–LOR architecture. The launch mode used in this architecture was given special attention because of the requirement for LVs differing significantly in size and capability from those in other architectures.

The EOR–LOR architecture uses one large LV to deliver the LSAM and EDS and one small LV to deliver the CEV into LEO, whereas the other architectures use a heavy-lift Cargo Launch Vehicle (CaLV) to launch the cargo elements and another heavy-lift Crew Launch Vehicle (CLV) to launch the CEV and crew. Due to the EOR–LOR architecture utilizing one large and one small LV, it is often referred to as a 1.5 launch mission. Although this architecture would appear to offer a streamlined alternative to the other architectures, the 1.5 launch configuration is only achieved after removal of supplemental radiation protection, which, in the eyes of the crew, represents a negative.

Transport

Another important aspect of architecture design was the ability to deliver large cargo elements to the lunar surface using the LSAM descent stage and EDS as one-way, unmanned transportation systems. The two architectures best suited to delivering cargo in a single launch were the LOR, capable of delivering 18 t, and the 1.5 launch EOR–LOR CaLV, capable of delivering 20.9 t. The reason for these impressive cargo loads is that in both of these architectures, the LSAM and EDS are launched together, with one of the elements performing LOI. The EOR–direct return, by comparison, is better suited for delivering cargo in two launches because in this architecture the LSAM is launched separately from the EDS. Thanks to this mission design, the EOR–direct return is capable of delivering 34.7 t to the lunar surface.

Analyzing the analysis

Having addressed all the myriad architecture configurations, the team performed a final analysis of mission mode viability by examining Figures of Merit (FOMs) discriminating between candidate architectures. Typical FOMs included measures such as architecture flexibility, reliability, P/LOC, P/LOM, safety, propulsion options, production costs, operations costs, and performance data specific to each mission mode.

Based on an assessment of FOMs the team were able to eliminate the EOR–direct return mission mode from further consideration due to the high number of operability problems, uncertainties, and poor safety margins inherent to the design. Another important outcome of the analysis was the decision of the study team to change the skin of the CEV from aluminum to a carbon composite due to the superior radiation-shielding properties of composite materials. Analyses performed on cost, risk, and performance values for each architecture revealed direct return missions exhibited the lowest development costs and the lowest risk of losing a crew but because of the operational complexity of the direct return mission mode, this architecture was ultimately rejected.

Based on these and other assessments, the study team prepared for Analysis Cycle 3 by focusing on a select number of studies they hoped would better define the 1.5 launch EOR–LOR architecture. The team also decided to refine the performance of the LOR and EOR–LOR launch modes as well as update the sizing of the LVs and investigate Lunar Lander and lunar surface configurations.

Mission mode evaluation, Analysis Cycle 3

The primary focus of Analysis Cycle 3 was a recommendation of a lunar mission mode based on the evaluations and studies conducted in the preceding analysis cycles. To that end the study team conducted further assessments directed at refining the lunar transportation system and CEV configuration, in addition to re-assessing the amount and type of supplemental radiation shielding. Another significant design hurdle that needed resolving was the issue of vehicle masses and the resulting performance margins associated with each mission component. Finally, once the team had performed all these evaluations and decided upon the lunar mission mode, it would be necessary to once again determine P/LOC and P/LOM.

Trade studies

In addition to the aforementioned assessments and evaluations the team was required to perform a series of trade studies designed to determine mission-critical factors such as radiation protection, risk leveling and emergency contingency procedures for a return from the lunar surface.

One of the trade studies refined the amount of radiation protection and the resultant risk associated with shield distribution and the addition of HDPE. Once the team had placed the shielding and constructed a high-fidelity model, they

calculated the radiobiological risk based on large SPEs, thereby enabling a calculation of the resulting probability of death and/or radiation sickness. The decision of
how much HPDE to use in the design of the vehicle was a critical one with far-
reaching consequences in terms of future crew selection. This was because those
astronauts with no prior occupational exposure would be within the NASA legal
dose limits if no HPDE was used, whereas astronauts with prior exposure would
exceed the 95% confidence limits and therefore be excluded from selection. However,
if the study team decided to use $2\,g/cm^2$ HPDE then some astronauts with prior
occupational exposure would be eligible for selection but others who had perhaps
served an ISS increment would be deselected.

The risk-leveling challenge of providing adequate radiation protection for crews
proved a tricky one since the team sought a solution producing a near-zero percent
probability of death or sickness, while also returning the crew to Earth without
violating career limits. After much number-crunching, the team calculated that by
using $2\,g/cm^2$ HDPE the probability of a death or acute radiation sickness would be
1 in 2,500 (0.04%) and career limits would not be exceeded. By removing all supplemental HPDE shielding, the probability of acute health effects was calculated to be
1 in every 1,428 (0.007%) missions. Given the extra mass of radiation shielding and
its performance impact upon the mission, the study team were forced to shed the
supplemental protection completely and instead recommended additional configuration studies aimed at reducing the radiation dose by optimal arrangement of crew,
fuel, and consumables. It was a decision guaranteed not to make any astronaut jump
up and down with joy!

Another trade study focused upon defining emergency procedures for leaving the
lunar surface in the event of an off-nominal event, such as an SPE or life-threatening
injury to one or more of the crew.

The study team wanted to provide astronauts on the lunar surface with an
"anytime return" option, giving the crew the ability to return from the Moon to
the Earth within five days of an emergency return declaration. To provide such an
option, the team had to calculate everything from worst-case lunar geometries and
inclinations of lunar parking orbits, to nominal transfer times from TLI to LOI.
Complicating the mathematical challenge was the requirement to calculate the
delta-Vs for each science/resource utilization site. Located at various latitudes, the
sites therefore required different delta-Vs, with polar and near-equatorial sites requiring the lowest delta-V and sites located near 75°N or 75°S and 25°E and 160°W
requiring the highest. The number-crunching headaches for the study team didn't
stop there because they also had to consider TEI opportunities which, if unavailable
at the time of the emergency return declaration, would require the crew to remain in
LLO for up to 14 days. Having an injured or radiation-impaired crew loitering in
LLO for 14 days is obviously not a palatable scenario since increased loiter time
results in increased risk to the crew with a concomitant increase in risk of LOC/LOM.
Unfortunately, the only way to avoid such a loitering event would be to provide extra
propellant to enable the crew to make a propulsive maneuver designed to place the
CEV in the correct orientation for TLI. Loath to add any more mass to the existing
design, the study team ultimately increased the lander mass and decreased the descent

stage, thereby enabling greater access to and from the lunar surface in the event of an emergency.

Now they had mostly resolved the radiation and lunar surface emergency problems, the study team turned their attention to developing a better understanding of the LSAM by conducting a configuration trade study.

The aim was to optimize the key functional capabilities of the LSAM, as well as to define the vehicle's design parameters as they related to specific mission modes such as the LOI maneuver, de-orbit, powered ascent, and terminal landing and rendezvous. In performing the LSAM study, the team studied every aspect of the design but paid particular attention to propellant types, hardware reuse, and airlocks. The study included a survey of previously considered vehicle concept designs, such as those investigated as part of the 2005 ESMD LSAM Phase 1 Study which was eventually used a starting point for the LSAM trade configuration analysis.

By studying various vehicle configurations, the team, utilizing a weighted scoring approach, were able to better understand the advantages and disadvantages afforded by each design and thereby select-in and select-out specific design and operational features. Ultimately, the team chose the two-stage configurations since these were assessed as providing optimal operational mission performance and lowest risk.

Having selected a configuration, the team now had to identify the elements that most impacted the vehicle performance in order to define key discriminators, allowing them to identify one of the two-stage options. Examples of some of the discriminators included payload unloading and whether to integrate the airlock within the CM or position it on another level (Figure 7.7).

One advantage of locating the airlock in the single-level habitation configuration was the quick access it provided the crew to the ascent stage in the event of an emergency ascent. However, in this configuration the CM and airlock location meant all surface sorties would begin at a greater height above the surface than in the split-level configuration, which provided the crew with easy access to the surface. Also, the split-level configuration afforded crews a higher degree of radiation protection due to the integration of the living quarters among the propellant tanks.

Having considered the airlock options, the team moved on to an evaluation of the propellant selection and its impact upon the LSAM configuration. Once again the team had to analyze several factors, such as whether to choose higher performance propellants such as hydrogen/oxygen at the expense of increased mission cost and increased volume, or to choose moderately performing propellants resulting in increased vehicle complexity. While the team wrestled with the propellant problem, other trade studies were being conducted to determine the optimal LSAM crew cabin configuration, resulting in three concepts submitted for consideration.

LSAM minimized ascent stage *configuration concept*

The focus of the first LSAM concept was to reduce the overall size of the ascent stage, requiring the study team to consider various permutations of crew positioning including all the crew standing, two crew standing and two crew sitting, and all the crew sitting. It was determined the all-crew-standing option would result in

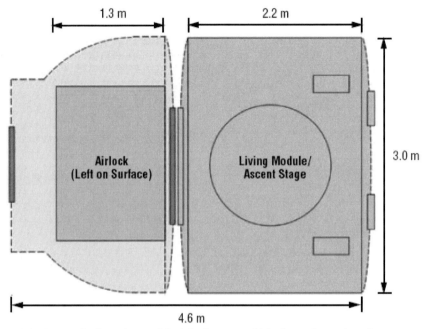

Figure 7.9. Lunar Surface Access Module Separate Airlock configuration. Image courtesy: NASA.

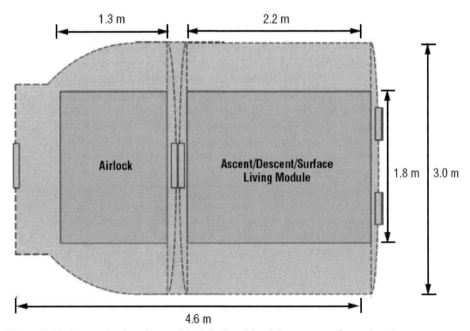

Figure 7.10. Lunar Surface Access Module Combined Concept configuration. Image courtesy: NASA.

Table 7.2. LSAM trade mass summaries [4].

	LSAM minimized ascent stage			LSAM separate airlock		LSAM combined	
	Ascent stage	Living Module	Descent stage	Ascent stage	Descent stage	Ascent stage	Descent stage
Dry mass	3,979	2,455	5,316	4,396	5,081	3,990	5,066
Inert mass	4,225	3,098	11,301	5,323	7,821	5,600	7,033
Total vehicle	7,811	3,153	41,348	9,570	35,937	10,203	34,985

ESAS LSAM final design

Once the study team had completed the LSAM studies the next step was to agree upon a design that would serve as a POD for future studies. After much additional analysis the team decided to go ahead with a combined design providing an airlock capability. What follows is a description of the elements of the reference LSAM concept for the ESAS 1.5 launch EOR–LOR architecture, a design bearing more than a passing resemblance to the Apollo LM configuration.

The LSAM ascent and descent stage will be capable of supporting four astronauts for seven days on the surface of the Moon and then ferrying the crew from the lunar surface to LLO. Utilizing an integrated pressure-fed oxygen/methane propulsion system, the ascent stage will perform ascent from the lunar surface to LLO, rendezvous, and dock with the CEV before separating and self-disposing. Maneuvering and attitude control will be achieved by 16 RCS thrusters, while the required 1,866 m/s ascent delta-V will be achieved using a single 44.5 kN propulsion system, drawing propellant from spherical propellant tanks (Figure 7.11, see color section).

The cylindrical LSAM will provide the crew with 31.8 m^3 of pressurized volume during lunar surface operations. After much deliberating, the study team decided upon a simultaneous EVA strategy requiring all crewmembers to egress the vehicle together, the same strategy employed by Apollo astronauts. In the present configuration the LSAM descent stage will perform the functions of inserting the CEV into LLO, landing the ascent stage on the lunar surface, and providing the vehicle's life support during the seven-day lunar surface stay. The descent will be achieved using a pressure-fed oxygen/hydrogen propulsion system comprising engines arranged symmetrically around the vehicle's centerline, as shown in Figure 7.11, color.

The eight propellant tanks mounted around the descent stage include six hydrogen and two oxygen tanks positioned in a circular arrangement, with two bays on opposite sides of the stage to permit surface access and cargo stowage. In this configuration, the descent stage also serves as the attachment point to the EDS, provides mounting locations for the active Thermal Control System radiators, and contains gaseous nitrogen and potable water required for the mission until the lunar ascent phase.

DEPLOYMENT ARCHITECTURE AND STRATEGY

Outpost architecture

The ESAS deployment architecture foresees the lunar outpost being completed in five deployment flights (Figure 7.12, see color section) but under this outpost construction strategy the five deployment flights will not be included in the lunar sortie mission phase. The goal of outpost completion in five flights also assumes precursor missions will accomplish tasks such as demonstrating ISRU technologies and establishing resources such as communication and flight navigation. If all requirements are met then the next goal of the outpost architecture will be to enable a continuous human presence, capable of sustaining an evolved set of surface capabilities. These will include frequent mid-field (<30 km) and far-field (>30 km) surface sorties, *in situ* data collection and analysis, regolith excavation and transportation, large-scale ISRU production, propellant production, and surface construction.

Outpost deployment strategy

The study team envisaged the lunar outpost being assembled and deployed piece by piece, utilizing an "incremental build strategy" where the outpost gradually evolves from hosting a semi-permanent crew to sustaining a permanent crew of four astronauts. The change from a semi-permanent to a permanent crew was envisaged to take three years or more and would be facilitated by the addition of necessary elements required to support permanent operations. These elements would be manifested into weight categories ranging from less than 2,000 kg to greater than 10,000 kg. For example, the deployment manifest for those elements weighing less than 2,000 kg would include equipment such as the unpressurized rovers (500 kg) and the ISRU Lunar Polar Resource Extractor vehicle (1,200 kg), whereas the manifest for elements weighing more than 10,000 kg would include equipment such as the Logistics Module and the Pressurized Rover, each of which weigh 10,000 kg. As with many of the evaluations of the study team a detailed assessment was recommended to determine whether to transport elements in modular pieces and build them on the surface or to transport them using dedicated cargo missions. Whereas it was relatively easy for the study team to envision piecing together a modularized Habitat Module, it was a little more difficult to envisage how astronauts might construct a complex element such as the Pressurized Rover.

Once the study team had assessed the various deployment options, they were able to define a manifest of the number and type of flights required to deploy the capabilities required for sustained lunar surface operations and for the incremental evolution of those operations.

Lunar architecture defined

Based on a comprehensive assessment of the findings of each analysis cycle and of numerous related studies, the study team finally decided upon an architecture considered both comprehensive and evolutionary. The architecture is capable not only of performing all the requirements of manned lunar missions but also of being modified for future manned missions to Mars and beyond.

The now-accepted VSE architecture, which will be a combined EOR–LOR, will be defined by the design of the CEV (Orion) and new LVs (Ares I and Ares V). The EOR–LOR will commence with the launch of Ares V, placing the LSAM and EDS in Earth orbit. Ares I will then launch, placing Orion and crew in Earth orbit, whereupon Orion and LSAM EDS configuration will rendezvous.

Once both launches have taken place and all mission elements are in Earth orbit the EDS will fire its two J-2S LOX/hydrogen engines to send the stack on a translunar trajectory, after which the EDS will be expended. The LSAM and Orion will then fly to the Moon and be inserted into lunar orbit by the descent stage of the lander. Once stable in lunar orbit, the crew will descend to the surface in the two-stage Lunar Lander, comprising a descent and ascent stage, leaving the unmanned Orion behind in orbit where it will operate autonomously during the crew's seven-day surface stay. After their lunar stay, the crew will fire up the ascent stage's pressure-fed LOX/methane engines and return to lunar orbit, dock with Orion, transfer back into Orion and depart lunar orbit using the SM propulsion system of the CEV. Finally, Orion will perform a direct Earth entry followed by a parachute water/land landing.

Surface architecture

The final architecture component describes how astronauts will spend their time while on the lunar surface. Defining this architecture was a little easier for the study team since the major details were all stated in the VSE, namely to develop and test new technologies that may be used for subsequent missions to Mars and beyond. To that end, the team decided the best way of accomplishing the VSE's goals was to focus on exploration science, development of lunar resources, and on evaluating operational techniques that may be used in future missions to Mars. A description of these activities is provided in Chapter 9.

RISK AND RELIABILITY ASSESSMENT

ESAS architecture risk and reliability assessment

As with all mission-planning and architecture development, the final element requires a risk assessment to ensure risks to the mission and crew are acceptable and also to confirm vehicle and mission requirements are met in terms of cost and performance. A risk analysis also provides mission planners with an opportunity to more carefully scrutinize those items assessed as high risk and to examine various configurations of mission elements in order to reduce any excess risk. Another goal of the mission planners was to identify and quantify architecture-discriminating risks that had the potential to compromise mission success, and then to resolve those risks in a way that did not adversely affect the overall architecture.

Table 7.3. Major lunar architecture risks [3].

1	LOX/methane engine development	6	Lunar vehicle LOX/hydrogen throttling on descent
2	Air start of the SSME	7	Integration of the booster for the HLV
3	Lunar–Earth re-entry risk	8	LADs in CEV propulsion system
4	Crew egress during launch	9	Unmanned CEV system in lunar orbit
5	J-S2 development for the EDS	10	Automated rendezvous and docking (AR&D)

Ultimately, the ESAS team sought to agree upon an architecture producing the highest probability of mission success, with the least risk to the safety of the crew. To that end analysis tools such as the Screening Program for Architecture Capability Evaluation (SPACE) and the Flight-oriented Integrated Reliability and Safety Tool (FIRST) were used to quantify factors such as mission risk and vehicle and mission element reliability.

Using SPACE and FIRST, the team determined a list of the major risks as shown in Table 7.3.

1. *LOX/methane engine development.* As alluded to previously, the problem associated with the LOX/methane engine was the simple fact this system was untested in flight and because it was not certain such a system could be flight-tested in time for the first launch date. However, because of its performance advantages, the system was considered worth pursuing, and ultimately the team determined that by pursuing an aggressive flight test schedule it would in fact be possible to meet the reliability criteria for such a system, and to do so in time for the first launch of Orion.

2. *Air start of the SSME.* Unlike the LOX/methane engine, the SSME is a flight-proven system with demonstrated high reliability. However, in the architecture agreed upon by the ESAS team, the SSME would be required to start up in flight, a scenario never before demonstrated on the ground or in flight. Due to the absence of data on SSME air starts it was decided to conduct an SSME test program. However, it was recognized that such a test program would need to be conducted on the ground and therefore could not exactly replicate the exact conditions at ignition. However, after further evaluation, and based on previous test experience, it was determined the system was likely to reach operational reliability in a timeframe consistent with the lunar architecture timeline.

3. *Lunar–Earth re-entry risk.* In assessing this risk, the primary focus of the team was on analysing the integrity of the Thermal Protection System (TPS). Although NASA had previous experience with ablative heat shields, developed for the Apollo program, Orion's size required a much larger area to be covered by ablative material. One option was simply to recertify existing Apollo material.

However, it was assessed that flight certification could be performed by using current computational fluid dynamics (CFD) technology and applying re-entry physics to perform high-fidelity simulations of Orion's TPS.

4. *Crew abort during launch.* Simulating a crew abort is enormously complicated, even with the aid of the most advanced CFD tools and other complex evaluation tests. Pressure and velocity profiles must be generated, fracture models mapped, internal and external engine conditions predicted, aerodynamic loads calculated, and even cloud interaction must be considered. Since crew abort requires the Launch Escape System (LES) to explosively separate from the vehicle, fracture fragment propagation in the air stream must also be analysed. Unsurprisingly, as this book is being written, NASA engineers are working to resolve this problem.

5. *J-S2 development for the EDS.* Since the J-2S engine has never been flown, the ESAS team were once again faced with the need to conduct a qualification, fabrication, and test program, which was estimated to take four years.

6. *Lunar Vehicle LOX/hydrogen throttling on descent.* As the Lunar Vehicle descends to the surface, it will be critical for the throttleable LOX/hydrogen engine to ensure injector pressure drops so that correct propellant injection and mixing occurs without any combustion instability. The limited throttling experience available to the team was from dual–throttling valves used on the Delta Clipper Experimental (DC-X) Program, which indicated throttling may be achieved using a sliding pintle design which would control the size of the engine orifice.

7. *Integration of the booster for the HLV.* The CaLV configuration which integrates two five-segment RSRBs with five SSME cores, uses mostly mature and reliable technology derived from the Space Shuttle. One of the main risks identified by the team was the uncertainty of adding a fifth segment to the SRB since the Shuttle has only four segments, and it was not known how this would affect thrust imbalances. However, given the heritage of the SRB it was concluded the integration issue was a manageable one.

8. *Liquid acquisition devices (LADs) in Orion propulsion system.* The main challenge associated with the design and flight ratification of the LADs was the problem of accurately modeling the temperatures and fluid properties. Fortunately, the LAD issue was another problem helped by the fact LADs are used in the Space Shuttle, so the team reviewed the history of the Shuttle LAD qualification program to determine the key issues.

9. *Unmanned Orion in lunar orbit.* Orion will remain unmanned for seven days in LLO, which means it must function flawlessly following its period of dormancy. However, any potential problems there may be with the quiescent period will most likely be resolved during early Orion flights to the ISS, when it will be dormant for periods longer than seven days.

10. *Automated rendezvous and docking.* The final choice of architecture does not involve AR&D, but AR&D will be required when pressurized cargo is delivered to the ISS. Automated docking carries a 1 in 100 probability of failure, as evidenced by events such as the Progress module impacting the Spektr module

of Mir in 1997, but team members rationalized that experience gained from early Orion missions would mitigate this risk.

Overall risk assessment from launch to landing

Based on the EOR–LOR architecture described previously, a risk assessment for each mission phase was performed. The risk assessment included reviews of previous missions such as Apollo, earlier risk assessments, analyses of maturity models for each mission element, consequence models to understand the impacts of risks upon achieving objectives, and the application of several other analytical tools designed to identify risk drivers and risk quantification. Since a complete review of the methodology employed to assess the risk of the architecture is beyond the scope of this publication, what follows is a synopsis of the assessed risks for each phase of the architecture, from the launch of the cargo and crew, to the entry and descent of the CEV, and landing back on Earth.

Launch Vehicle risk

More than 30 LV concepts were evaluated by the Marshall Spaceflight Center (MSFC) Safety Mission Assurance Office (S&MA) before choosing the CLV LEO Launch System most suitable for the EOR–LOR architecture. By applying LV reliability estimates, vetted failure rates, and calculating probabilities for each system and subsystem of each LV configuration, the S&MA was able to calculate the probability of LOC as 1 in 1,429 and the probability of LOM as 1 in 182.

A similar process was performed in the assessment of the reliability of the in-space liquid propulsion system. In order to accurately assess this risk, it was necessary to identify the components likely to have the highest impact upon reliability. By using modeling of the physical elements that comprise the propulsion system, it was possible to identify high-risk elements such as the Engine Purge System, which plays an essential role in the restart. Once again, previous experience with SSMEs helped the team analyze the risk associated with key elements, such as isolation valves.

CEV, SM, and LSAM risk

Since the CEV, SM, and LSAM had no heritage it was not possible to develop exact failure rates for every system, although failure rates for subsystems from which information could be derived from other space subsystems were estimated. In this probability estimate of mission elements, the team were focused on calculating an aggregation estimate of catastrophic failure for the CEV, SM, and LSAM mission phases so that levels of redundancy could be determined and appropriate trouble-shooting solutions could be designed to increase the chance of mission success in the event of a failure of one or more subsystems.

Rendezvous and docking hazards

Rendezvous and docking maneuvers will be conducted in LEO or LLO. For each of these maneuvers different contingency and malfunction procedures will be required. Fortunately, NASA has achieved a 100% rendezvous-and-docking success rate in its missions to date, thanks to several levels of subsystem redundancy and well-established rendezvous techniques.

The rendezvous will be achieved using the same equipment fitted to the Space Shuttle, such as the Ku-band antenna used in radar mode to track the target vehicle. If a Ku-band antenna fails, as occurred during STS-92, secondary rendezvous equipment such as the Star Tracker can be called upon, and even if this fails, a tertiary level of redundancy in the form of a hand-held light detection and ranging (LIDAR) system can be used. Based on 113 Shuttle flights, it is calculated the chances of all proximity operation instruments such as the Ku-band and Star Tracker failing is an extremely low 6.4^{-7}. The most likely equipment malfunction that may jeopardize a rendezvous is a failure of the RCS jet to burn when commanded but, since Orion has several jets in each axis, the RCS has a high level of redundancy. In fact, in an evaluation of 113 Shuttle flights, it was calculated the probability for all RCS jets to fail was 3.8^{-6}.

Lunar surface stay risk

The main risk driver element while the crew is on the lunar surface is the reliability of the LSAM, which must power down once the crew departs for their seven-day surface stay. Since there was no history of the LSAM or its systems, the risk was calculated by summing the probability of failure per hour of each system and subsystem involved, and multiplying the resultant summation by 24 to provide a daily quiescent failure rate of the LSAM as a whole. Based on these calculations it was determined the LSAM failure rate probability was 4.86^{-5} per hour and the corresponding failure rate for the CEV/SM was 6.51^{-1} per hour. Summing the two failure rates and multiplying by 24 results in a 2.73^{-1} per day risk of failure.

By pursuing an exhaustive process of trade studies and analysis, NASA has defined a versatile and mature lunar architecture enabling the proving of technologies for future manned missions to Mars. However, there are some within the space community who consider abandoning the current architecture, fearing a manned Moon base will bog down the space program for decades and inhibit, rather than facilitate, manned Mars operations.

REFERENCES

[1] *America at the Threshold: Report of the Synthesis Group on America's Space Exploration Initiative*. U.S. Government Printing Office, Washington, D.C. (May 1991).
[2] Cohen, A. *Report of the 90-Day Study on Human Exploration of the Moon and Mars*. NASA Johnson Space Center, Houston, TX (November 1989).

[3] *http://www.nasa.gov/mission_pages/exploration/news/ESAS_report.html*

[4] *http://www.nasa.gov/directorates/esmd/home/index.html*

[5] NASA. *First Lunar Outpost Requirements and Guidelines (FLORG) Fully Annotated Working Draft*, EXPO-T1-920001. NASA Exploration Programs Office, Johnson Space Center, Houston, TX (June 10, 1992).

[6] NASA. *Human Lunar Return Study: Status Review Continuation*. NASA Johnson Space Center, Houston, TX (June 17, 1996).

[7] Ride, S.K. *Leadership and America's Future in Space*, a report to the Administrator of NASA. NASA, Washington, D.C. (1987).

[8] Roberts, B.B.; Bland, D. *OEXP Exploration Studies Technical Report*, Vol. 3, NASA-N-89-15845, NASA-TM-4075. NASA, Washington, D.C. (December 1988).

8

Alternative mission architectures

*The urge to explore has propelled evolution since the first water creatures
reconnoitered the land. Like all living systems, cultures cannot remain static;
they evolve or decline. They explore or expire. Beyond all rationales, spaceflight is a
spiritual quest in the broadest sense, one promising a revitalization of humanity and
a rebirth of hope no less profound than the great opening out of mind and spirit at
the dawn of our modern age.*

Buzz Aldrin, "From the moon to the Millenium," *Albuquerque Tribune*, 1999

As this book is being written, a U.S. presidential election is looming and some space
leaders are working to offer the next U.S. president an alternative to President Bush's
"Vision for Space Exploration" (VSE)/Constellation Program [4]. Rather than return
to the Moon, this influential group of leaders propose moving towards manned
missions to asteroids, arguing such a strategy would lead to earlier manned flights
to Mars. Alternative strategies to Constellation include missions sending astronauts
to Lagrangian points, 1.6 million kilometers from Earth, where the gravity of the
Earth and Sun cancel each other out. The alternative vision planners, along with
others in the scientific community and some of the general public, believe the VSE is a
step backwards and will only serve to decelerate manned spaceflight operations
beyond the Moon. It is important therefore to examine the reasons for returning
astronauts to the lunar surface, just as it is essential to also address the issues raised
by the alternative vision planners.

THE MOON AS A PROVING GROUND: VSE VS. THE
ALTERNATIVE VISION

One of the main arguments made by the VSE is using the Moon as a proving ground
for future exploration-class missions to Mars. The myriad challenges of a manned

Mars mission can be classified into seven groups: human health and performance, life support systems, mobility systems, system reliability, dust mitigation, transportation systems, and autonomous operations. These challenges are addressed here in the context of the current VSE and the claims made by alternative vision planners.

Human health and performance

Human health and performance factors include the development of radiation environment and protection strategies, the definition of extended-duration counter-measure protocols, medical diagnosis equipment, and human machine efficiency. VSE proponents argue that by returning to the Moon it will be possible to validate the efficacy and performance of countermeasure equipment, test and demonstrate medical equipment, and validate human factors. Perhaps most importantly, extended-duration Moon missions will provide an opportunity to accurately define the radiation environment and permit the calculation of interplanetary radiation thresholds for Mars-bound astronauts. Given the potential impact of radiation on interplanetary crews, and the absence of NASA-defined radiation threshold limits, the argument for returning to the Moon, as opposed to heading directly to Mars, is a compelling one and would be reason enough to return to the Moon.

Life support systems

Life support system (LSS) challenges for a Mars mission include the development of advanced closed-loop support systems that must function with high reliability and maintainability. Once again, the VSE group argues that long-duration testing with humans while on the lunar surface provides the perfect opportunity to understand optimal system performance and crew operation. Such an argument would make sense since it is better to resolve LSS issues at a location four days flying time away than encounter a problem *en route* to or on the surface of Mars.

Mobility systems

Each mission to Mars will require extended surface operations which can only be sustained by advanced mobility systems capabilities. While the integral mobility systems component is the space suit, associated systems such as life support, radiation protection, and extravehicular activity (EVA) system maintenance strategies must also be developed. Since the only local planetary venue where such systems can be validated is the Moon, it is difficult to argue against the lunar surface as a proving ground.

System reliability

A Mars crew will be away from home for at least two years, making system reliability and maintenance critical issues. To best support such a mission, it is critical to develop repair and maintenance facilities, crew autonomy and training concepts, and integrated logistics support. These issues must be refined and the systems must be robust before embarking upon a manned mission to Mars. Again, the Moon provides the perfect test-bed for the development and testing of these facilities and concepts.

Dust mitigation

Each Mars mission component and system will be exposed to planetary surface dust for extended periods. EVA systems, planetary rovers, electronic components, and lander systems will need to be protected by effective dust mitigation techniques and operational procedures designed to reduce the insidious effects of dust accumulation and penetration. The lunar surface, although not truly "Mars-like", is the only means by which dust mitigation strategies can be accurately tested. Embarking upon a manned mission to Mars before resolving the dust problem that caused the Apollo astronauts so many problems would surely be constituted by many as an abdication of responsibility on the part of mission planners and would unnecessarily increase the risk of what is already a risky mission.

Transportation systems

Automated rendezvous and docking (AR&D), Earth return at high speeds, and advanced cryogenic propulsion represent some of the transportation system challenges that must be resolved before sending a crew to Mars. AR&D maneuvers are well proven but flight experience conducting these operations with minimal Earth support is limited, and gaining such experience does not require traveling to the Moon. Equally, on the subject of developing cryogenic fluid storage and management systems, such a process can be achieved just as effectively during missions to and from asteroids or to L1 points. In the case of transportation systems the alternative vision planners have a case.

Autonomous operations

The challenge posed by autonomous operations is determined by the requirement to develop highly reliable software verification systems, model-based reasoning (MBR) techniques, and fault detection methods. While the alternative vision planners might argue that developing these systems may just as easily be achieved in their revised VSE, the reality is the Moon provides testing in actual operational conditions and circumstances more analogous to the surface of Mars.

Reinventing Apollo

Alternative vision planners argue a return to the Moon is merely an exercise in reinventing Apollo, whereas missions to the Earth–Sun Lagrangian points, the first excursion to the limit of Earth's influence, would represent a significant step beyond Apollo. In addition to visiting the Lagrangian points, alternative vision planners suggest landing on an asteroid, designated *99-A010*, using a plan developed by planetary scientist Robert Farquhar. The five-month mission, launching in 2025, would use an Ares V booster to send astronauts to the asteroid for a 30-day stay. Accommodation would be courtesy of Bigelow Aerospace, which would provide one of their pressurized modules to house the crew during surface science operations.

The flight would demonstrate the feasibility of a scaled-down version of a mission to land on the Martian moons Phobos or Deimos. However, the intent of the VSE is to land on Mars, and spending time on an asteroid does not help program managers resolve many of the aforementioned "manned mission to Mars" challenges [4]. Furthermore, alternative vision planners argue an asteroid mission would accelerate manned missions to Mars, but if the first asteroid mission were to take place in 2025, it is unlikely a Mars mission would take place much before 2030, the date mandated by the original VSE. Finally, although sending astronauts to asteroids as opposed to *Apollo on steroids* may reduce cost by not having to build lunar surface systems, the alternative vision group's plan still requires extra hardware. For example, one alternative to VSE calls for a Deep-Space Shuttle (DSS), comprising a Service Module with a chemical propulsion system and crew quarters, a detachable, Apollo-style Re-entry Vehicle, and an Interplanetary Transfer Vehicle (ITV). Although the Crew Exploration Vehicle (CEV)/Orion currently under development by NASA could serve as the re-entry vehicle, the DSS and the ITV would still require development.

The inevitability of VSE

Arguments for and against "manned lunar missions" vs. "astronauts to asteroids" aside, the reality is, regardless of the advantages and disadvantages, the manned mission to the Moon is moving ahead, as evidenced by hardware being built and contracts being awarded. However, as with all major enterprises, the VSE is suffering from the inevitable problems of budget and scheduling, a situation compounded by the necessity to retire the Shuttle in 2010.

> "The FY07 appropriations, if enacted as the House has resolved, will jeopardize our ability to transition safely and effectively from the Shuttle to the Orion Crew Exploration Vehicle and the Ares I Crew Launch Vehicle. It will have serious effects on people, projects, and programs this year and for the longer term."
>
> Dr. Michael D. Griffin, NASA Administrator, February 5, 2007
> (Press Conference at NASA HQ)

By NASA's own admission, the 2007 budget has increased the manned spaceflight capability for each of the next five years, if the agency remains committed to the mission architecture outlined in the Exploration System Architecture Study (ESAS) [6]. Once the Space Shuttle is retired in 2010, NASA will lose independent access to the International Space Station (ISS) and will instead be reliant upon Russian Soyuz flights to ferry astronauts to and from the ISS, a situation provoking the NASA administrator to voice his frustration:

"I think personally that it is unseemly for the greatest nation in the world, today's pre-eminent space-faring nation to be in a position where we have no other choice but to buy rides from Russia."

Dr. Michael D. Griffin, NASA Administrator, June 8, 2007
(interview with Reuters)

The rapidly increasing gap between the Shuttle's retirement and the beginning of Ares I/Orion operations is not the only concern voiced by those in the space industry since, by committing to the current VSE, NASA will lose its STS-based infrastructure and workforce. The irony of such a situation is not lost on those who remember the dismantling of the infrastructure that sustained the Apollo program, nor is such an unnecessary fate far from the mind of one of the few people with the ability to remedy the current state of affairs.

"From 1975–1981, between the retirement of the Apollo–Saturn system and the first flight of the Shuttle, the United States did not have the capability to send humans into space, our country was not driving the space exploration agenda, and our aerospace workforce was decimated. We lost valuable people from the program, people who never came back. We lost valuable skills that were relearned with difficulty, or not at all. We lost momentum. Let us learn from these experiences. Let us not repeat them. Let us at least make a new mistake."

Dr. Michael D. Griffin, NASA Administrator, August 31, 2005
(AIAA Space 2005 Conference and Exhibition)

In light of the above arguments and with an impending change of administration looming, it is appropriate to examine an alternative VSE. Although a number of alternatives to VSE have been suggested, the most prominent is the one known as DIRECT.

DIRECT

In October 2006, an alternative to Constellation was published by a team led by Ross Tierney, a British citizen and CEO of Launch Complex Models. Known as the Direct Shuttle Derivative, or DIRECT, the study was designed to persuade NASA to consider replacing the planned Ares I and Ares V with a single Launch Vehicle (LV) developed directly from existing Space Shuttle components [3]. The DIRECT LV would be based entirely on existing Space Shuttle hardware and infrastructure and designed to be configured in several different ways to match specific mission profiles.

On October 25th, 2006 the study was submitted to NASA Administrator Michael Griffin, and shortly thereafter, Dr. Doug Stanley, author of NASA's ESAS Report, provided a critique, resulting in the re-evaluation of the entire proposal in a refine-ment study. After several months of revised calculations, feedback, and critical analysis, Tierney and his team, which included many engineers and mid-level managers with NASA, published what is now known as DIRECT v2.0 on May 10th [3], and presented it on September 19th, 2007 at the American Institute of Aeronautics and Astronautics (AIAA) *Space 2007* conference in Long Beach, California. If adopted, DIRECT v2.0 promises to save NASA $35 billion over the next 20 years. This chapter describes and evaluates Tierney's proposal.

DIRECT v2.0 drivers

The starting point for the DIRECT v2.0 (which is referred to simply as "DIRECT" in this chapter) team was to identify a new LV design based on a number of driving imperatives, the first of which was to shorten the period during which the U.S. would be totally dependent on Russian Soyuz spacecraft to ferry American astronauts to the ISS. The second imperative was to provide an LV with a launch capability to LEO greater than that of the Space Shuttle. The third imperative was to avoid duplication of existing U.S. Launch Vehicles, such as the Lockheed Martin and Boeing Evolved Expendable Launch Vehicle (EELV) fleet. Finally, to retain the current launch infrastructure, the DIRECT team's fourth imperative was to ensure any new launch system would be Shuttle-derived, so that much of the existing infrastructure could be used.

DIRECT history

The public face of the team proposing DIRECT includes Steve and Philip Metschan, Antonio Maia, Chuck Longton, and Ross Tierney. The private face comprises nearly 50 engineers, analysts, and managers, most of whom work for NASA HQ and its field centers. The DIRECT concept is not new. A similar idea was proposed by NASA's Marshall Space Flight Center (MSFC) in the aftermath of the Space Shuttle Challenger tragedy, but Congress informed NASA there were no funds available for building new LVs, so NASA decided to fix the Shuttle instead. Less than five years later, a NASA and Department of Defense team suggested a similar system based on the design of the Shuttle's solid rocket boosters (SRBs), but once again NASA was told no money was available. Another design similar to DIRECT was suggested by the ESAS team in 2005, which used the Shuttle SRBs, a Core Stage based on an existing Shuttle External Tank (ET), and three Space Shuttle Main Engines (SSMEs). The LV was designated LV-25 in its cargo configuration and LV-24 in its crew configuration. Although the LV had the capability of using an Earth Departure Stage (EDS) similar to the Ares V configuration the full extent of the design was never fully examined by the ESAS team. However, independent analysis of the LV-25 configuration indicated that, with an EDS, the vehicle could have launched 110 tonnes to LEO. ESAS discounted the LV-25/EDS configuration on the assumption only an Ares I LV could fly the CEV/Orion. This assumption resulted in another two launches being required to send other lunar payload components into orbit. Since such an architecture would require three launches according to the ESAS team, the LV-24/LV-25 vehicle was taken off the drawing board. This was unfortunate since, upon further independent assessment of the LV-24/LV-25 vehicle, it was determined its launch capabilities comfortably fit into a two-launch architecture and it exceeded the crew safety requirements and payload performance stipulated by ESAS.

Jupiter Launch System

Jupiter 120 and Jupiter 232 overview

The hardware (Figures 8.1 and 8.2) for DIRECT's Jupiter 120 and Jupiter 232 Launch Vehicles is comprised of many elements familiar to those who have followed the Space Shuttle missions over the last three decades (Table 8.1).

The Jupiter Launch Vehicles

The elements depicted in Figures 8.1 and 8.2 and described in Table 8.1 constitute the Jupiter Launch System (JLS). The crew variant of the Jupiter 120 comprises one cryogenic stage, two main engines, and no upper stage and will be capable of delivering 45 tonnes and Orion, whereas the cargo variant will be capable of launching 48 tonnes into LEO. This launch capability compares with the Ares I lift capacity of 22 tonnes into LEO, effectively duplicating the performance of existing LVs such as the Delta IV Heavy and Atlas V.

The second LV, designated Jupiter 232 (Figure 8.3, see color section), consists of two cryogenic stages, the first fitted with three engines, and the upper with two engines. This LV is capable of delivering 108 tonnes to LEO. The naming convention

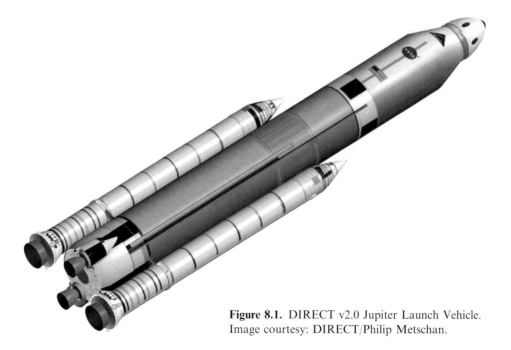

Figure 8.1. DIRECT v2.0 Jupiter Launch Vehicle. Image courtesy: DIRECT/Philip Metschan.

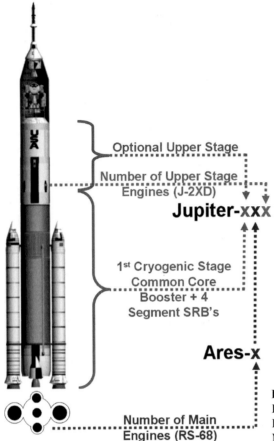

Optional Upper Stage

Number of Upper Stage
Engines (J-2XD)

Jupiter-xxx

1st Cryogenic Stage
Common Core
Booster + 4
Segment SRB's

Ares-x

Number of Main
Engines (RS-68)

Figure 8.2. Components of the DIRECT Jupiter Launch Vehicle. Image courtesy: DIRECT/Philip Metschan.

using a three-digit designator is used to identify the number of cryogenic stages used prior to Earth orbit insertion (EOI) (first digit), the number of main engines on the core (second digit), and the number of engines fitted to an US (third digit). Since the Jupiter 120 does not have any engines fitted to an US, the third digit is zero.

Payload

The ESAS plan calls for a capability of delivering 150 tonnes to LEO in two launches, a payload many space observers consider marginal to ensure the safety of four astronauts during a lunar mission. The 150 tonnes of initial mass in low Earth orbit (IMLEO) requirement is not a concern for DIRECT planners since the two-launch Jupiter 232 is capable of delivering 95 tonnes to low Earth orbit (LEO) *per flight*, thereby exceeding the ESAS IMLEO requirement by 40 tonnes!

Table 8.1. DIRECT Jupiter 120 and Jupiter 232 Launch Vehicle concept specifications [3].

	Jupiter 120		*Jupiter 232*
GLOW	2,033,940 kg	**GLOW**	2,339,490 kg
CLV Launch Abort System mass	6,565 kg	CaLV aero fairing mass	2,279 kg
Booster		**Booster**	
Propellants	PBAN	Propellants	PBAN
Usable propellant	501,467 kg	Usable propellant	501,567 kg
# Boosters/Type	Two/Four-segment Shuttle RSRMs	# Boosters/Type	Two/Four-segment Shuttle RSRMs
Core stage		**Core stage**	
Propellants	LOX/LH$_2$	Propellants	LOX/LH$_2$
Usable propellant	728,002 kg	Usable propellant	729,002 kg
# Engines/Type	Two/RS-68	# Engines/Type	Three/RS-68
Engine thrust @ 100%	SL[a] 297,557 kgf Vac[b] 340,648 kgf	Engine thrust @ 100%	SL 297,557 kgf Vac 340,648 kgf
Core burn time	446.0 s	Core burn time	292.0 s
LEO delivery orbit	77.5 × 222.2 km @ 28.50	**Second-stage EDS**	
Maximum payload (gross)	46,635 kg	Propellants	LOX/LH$_2$
Maximum payload (net)	41,971 kg	Usable ascent propellant	216,012 kg
		# Engines/Type	Two/J-2XD
		Engine thrust @ 100%	124,057 kgf
		Ascent burn time	392.0 s 55.6 × 222.2 km @ 28.50
		Maximum payload (gross)	105,895 kg
		Maximum payload (net)	95,305 kg

[a] Sea level.
[b] Vacuum.

Integration and utilization of Shuttle-derived technology

Solid rocket boosters

First, DIRECT intends to use the Space Shuttle four-segment SRBs which, following their redesign in the wake of the Challenger disaster in January 1986 [1], have performed flawlessly (Figure 8.4). Comprising a motor, a separation system, flight instrumentation, recovery avionics, pyrotechnics, parachutes, a Thrust Vector Control System and a range safety destruct system, the SRBs mode of operation is relatively simple. The SRB propellant consists of an oxidizer (ammonium perchlorate), fuel (aluminum), and a binder. Once the SRB is ignited it will continue to burn until the pressure inside the chamber reaches a prescribed level, resulting in the SRB's separation from the vehicle by means of a NASA Standard Detonator (NSD). Once all fuel is spent, the SRBs freefall back into the atmosphere, encountering aero-thermodynamic heating, until at an altitude of 5 km, parachutes deploy and the SRBs land in the ocean. Following dismantling and assessment, the SRBs are returned to the manufacturer for cleaning, inspection, and repacking with propellant for reuse on subsequent missions.

Each of the events described here has been successfully demonstrated numerous times on STS missions. Since the SRBs are existing, human-rated elements there is no

Figure 8.4. Space Shuttle solid rocket booster. Image courtesy: NASA.

cost or schedule impact as a result of development, and because there are sufficient SRB components to support more than a hundred Jupiter missions, there are also no costs involved. The ESAS architecture also intends to use the SRBs as a part of the ALS but, instead of a four-segment SRB, the ESAS plan requires the development of a five-segment version using a different fuel formulation. Developing a five-segment SRB may sound fairly simple, given the booster's relatively straightforward operation, but the reality is such a requirement demands modifications to the SRB infrastructure, such as recovery ships, work platforms inside the Vehicle Assembly Building (VAB), and the Mobile Launcher Platform (MLP).

External Tank

For the Jupiter Common Core Booster (CCB) DIRECT decided to use the Space Shuttle's ET. The CCB represents the principal element with which the other JLS components will be integrated and will serve as the building block for subsequent derivatives of the JLS. Once again, DIRECT avoids the disruption to production, transportation, and workforce infrastructure by deciding to use the ET in its current configuration, rather than build a different one as mandated by the ESAS plan.

The ET comprises a forward liquid-oxygen cryogenic tank, constituting more than 80% of the total mass of the ET, an intertank, and an aft liquid-hydrogen cryogenic tank. Liquid oxygen (LOX) is delivered to the SSMEs via an externally mounted pipe running outside the LH_2 tank, separated from the LOX tank by the intertank, whose function is to receive and distribute loads between the two tanks.

The LH_2 tank is the largest cryogenic tank by volume, although its fully loaded mass is just one-fifth of the LOX tank, since the density of LH_2 is so much lighter than LOX. A feed line jutting out from the base of the tank delivers LH_2 to the SSMEs.

To become the Jupiter CCB, the ET will require modifications such as strengthening the sidewalls, designing a new avionics system, and developing some new plumbing, all tasks within the capabilities of the ET production facility based at the Michoud Assembly Facility (MAF) in New Orleans, Louisiana. For example, the pointed nose of the current ET will need to be altered to a dome shape, permitting a payload interface capable of transmitting rotational and axial loads to the forward-mounted elements. Some alterations will also be required in the ET's cryogenic plumbing, specifically the pipes leading from the LOX and LH_2 tanks, in order to increase flow rates.

Integration and utilization of existing technology

DIRECT does not use the SSMEs, arguing that since the engines must be replaced after 20 uses, it doesn't make any sense to use them on a disposable Jupiter Common Core (JCC). Instead, DIRECT identified the Pratt & Whitney Rocketdyne RS-68 engine (Figure 8.5), originally designed for the U.S. Air Force's Delta IV program. Inexpensive enough to be disposable and more than one and a half times as powerful

Figure 8.5. RS-68 rocket engine on Stennis Space Center's A-1 Test Stand. Image courtesy: NASA.

as the SSME, the RS-68 requires only human-rating before being qualified for flight, thereby removing any significant development costs.

The RS-68 rocket engine

One of the few new hardware elements required by the Jupiter Launch System (JLS) is the Aft Thrust Structure (ATS) of the CCB, providing the means by which the

RS-68 engine is attached. The RS-68 engines are currently being used on the Delta IV Evolved Expendable Launch Vehicle (EELV) and represent the first new American booster engine developed in over a quarter of a century. The end result of the RS-68 development was an engine generating 50% more thrust than the SSME and requiring 80% fewer parts to build, resulting in significantly reduced production costs.

How the RS-68 rocket engine works

A primary characteristic of the RS-68 engine is the use of a gas generator cycle with two independent turbopumps. The gas generator produces hot gas by tapping off a small percentage of the propellant flow. The gas is then sent through a turbine, which produces the pressure required to drive propellants into the main combustion chamber, after which the hot gas is expelled from the turbine.

By increasing propellant flow into the gas generator, the turbine speed is also increased, which increases propellant flow into the main combustion chamber, thereby increasing thrust. One of the few performance decrements associated with the design of the RS-68 is the reduction in specific impulse, caused by the need for the gas generator to burn propellants at a less efficient fuel-to-oxidizer ratio in order to maintain gas temperature that doesn't harm the turbine blades. Unlike the SSME, which incorporates a tube-wall design, the RS-68 features a channel-wall combustion chamber made from an ablative material which, although heavier than the tube-wall design, is simpler and less expensive to manufacture. Another advantage of the RS-68 engine is that the Jupiter 120 doesn't require the engine's full capability for the first crewed missions to the ISS, which means development to increase efficiency and thrust, as demanded by the ESAS plan, is not required. Also, since the RS-68 would be operated well within performance limits, the DIRECT plan does not run the risks of the current Ares I architecture requiring engines to be operated close to their rated performance limits after only a few test flights.

Jupiter Launch System Upper Stage

To enable a two-launch architecture for lunar missions, DIRECT would add a JLS US, but this element would not be required until after the missions designated to support the ISS. Such an approach means that while the Jupiter 120 flies crews to and from the ISS, the JLS can be developed over a period of between four and five years. Such a plan will not result in the rushed approach required by the ESAS recommendations. The components of the JLS US are described here.

Long-duration cryogenic tank

A key element of the JLS US is a long-duration cryogenic tank, since one of the requirements supporting long-duration missions is the storage of extremely low temperature cryogens such as LH_2 and LOX for long periods of time. The requirement for storing LH_2 and LOX is a function not only of expected mission duration but also of contingency planning. Such a contingency would occur in the event of an emergency on the lunar surface, requiring the crew to stay longer than planned, or in the event of a missed rendezvous.

Rocket engines

The JLS US will use Rocketdyne's J-2X vacuum-optimized engine, a derivative of the massive J-2, the largest upper-stage liquid hydrogen–fueled rocket engine ever developed by the United States. Five J-2 engines were used in the second stage of the Saturn V configuration and one for the EDS but, although the size of the engine was impressive, its most distinguishing feature was its re-start capability following a shutdown.

The J-2XD, a derivative of the J-2S, which, in turn, evolved from the J-2, would, in common with the RS-68, be designed with simplicity and reliability in mind, by reducing parts and improving engine start-up reliability. The engine start-up cycle of the J-2S was kicked off by a solid-gas generator which ensured the oxidizer and fuel pumps were primed. To provide engine re-start capability, solid propellant cartridges were stored within the manifold and, because the engine could be throttled, operating pressures were optimized using a variable mixture system. The J-2 variants are compared in Table 8.2.

The J-2XD used by the JLS US is a direct derivative of the J-2 and does not require any further development, whereas under the ESAS plan a new higher flow turbopump is required to improve performance. In fact, the only difference between the J-2 design and the J-2XD is the nozzle extension, and a slightly modified turbopump design was recently used in the linear aerospike X-33 engine development program. So, whereas the ESAS plan requires an aggressive development of a more advanced engine than the J-2, the DIRECT plan already has an engine ready to go.

Table 8.2. J-2 variants: JLS Upper Stage engine specifications [3].

	Apollo		*Jupiter*	*Ares I and V*
Upper Stage engine options	J-2	J-2S	**J-2XD**	J-2X
Turbopump	J-2 heritage	J-2 heritage	New	New
Nozzle	J-2 heritage	J-2 heritage	J-2 heritage	New
I_{SP} (vacuum), s	425	436	**448**	448
Thrust (vacuum), N	1,023,091	1,179,000	**1,217,000**	1,307,777
Flow rate, kg/s	245	276	**277**	298
Expansion ratio	27.5:1	40:1	**80:1**	80:1
Length (m)	2.95	2.95	**4.37**	4.37
Diameter (m)	2.03	2.03	**2.85**	2.85

Interstage

An interstage was a prominent feature of the Saturn Launch System (SLS) and the interstage separation event was one of the most memorable phases of the Saturn V's launch to orbit. The interstage element of the JLS features solid rocket motors which will provide a kick to the separation between stages and prevent collision between flight components.

Jupiter Launch System vehicle launch infrastructure

Assembly and processing infrastructure

Since the DIRECT architecture is Shuttle-derived, the transportation and processing equipment used for the Shuttle, such as the CCB, would also be suitable for the JLS components with few modifications required. For example, the Pegasus transport barge carrying the ET element of the STS would still be used to carry the JLS CCB. The ESAS architecture, however, would require the barge to undergo extensive alterations in order to transport the Ares I Upper Stage (US) and even more alterations to transport the Ares V. Obviously, such modifications would require costly and disruptive changes at MAF, a change in transport infrastructure, and changes to the production line. Given the high degree of commonality between the DIRECT elements and existing infrastructure, none of these disruptions would result.

The flight-ready CCB would be assembled at MAF in much the same way as the Saturn V's first stage was processed for the Apollo program. This architecture feature would ensure only activities directed at the final integration and assembly of the SRB Jupiter Upper Stage (JUS) would be performed at KSC.

Launch infrastructure

The JLS launch infrastructure comprises an 80 m tall MLUT which will hold the umbilical hook-ups for the Jupiter Upper EDS and mission components, such as the Orion and Lunar Surface Access Module (LSAM)/Altair. The 103 m permanently fixed FSS, requiring the addition of four stories and possible extra bracing, will be used only for maintenance access for the taller vehicle configurations by the use of five 22 m service arms.

DIRECT vs. ESAS launch facilities

The key to the DIRECT architecture is maximizing the use of existing Shuttle infrastructure. In sharp contrast with the infrastructure requirements for the Jupiter LV family, Ares I will require building new MLPs, a new Launcher Umbilical Tower (LUT), and extensive refurbishment of the work and service platforms inside the VAB. Furthermore, Ares I will require disposal of existing launch tower facilities such as the Fixed Service Structure and Rotating Service Structure (FSS and RSS), while Ares V will require new crawlers due to the mass of the five-segment SRBs and attached LUT on the MLP.

Since Ares I and Ares V are such different vehicles, the ESAS plan demands NASA operate two facilities to service what in effect are two separate and radically different launch systems. Although NASA has indicated it intends to solve the problem with two launch systems by implementing a "clean pad" concept, in reality, due to the divergent launch systems, this approach may not be feasible. Basically, the ESAS architecture infrastructure has little or no commonality, a feature likely to cause problems that will accumulate as radically different launch systems are operated.

In contrast, the modifications demanded by the JLS are almost non-existent thanks to the majority of the flight components being derived directly from existing hardware. The JLS will require some minor modifications to the MLP such as reconfiguring the Water Suppression System plumbing and the use of a Minimal Launcher Umbilical Tower (MLUT). However, these costs are negligible compared with those involved in the ESAS plan.

The lean and simple tactic of the DIRECT plan, combined with the utilization of a clean pad approach, means the JLS can launch back-to-back with a turnaround of less than ten days. Not only is such a fast turnaround a huge advantage when the unpredictability of the Florida weather is considered, but such a capability has a significant influence upon launch windows and reducing the probability of LOM. These latter benefits become even more critical when future manned missions to Mars are considered due to the requirement for multiple launches.

Trade studies

The JLS plan was subjected to several analytical simulations designed to determine safety margins and performance limits, reduce development costs, and optimize launch configurations for different tasks. For example, an assessment of the optimum mission architecture was studied to determine whether the most advantageous mission profile was to use Earth orbit rendezvous (EOR), lunar orbit rendezvous (LOR), or and Earth–Moon Lagrange rendezvous (EMLR).

Safety analysis

Space Shuttle lessons

One of the problems inherent in the Space Shuttle design is the requirement for the mission payload to be carried internally by the crew vehicle. Such a configuration increases mass, a feature compounded in the Shuttle design by the requirement of reusing the SSMEs, increasing vehicle mass even more. The mass increase has a direct impact upon the crew's safety envelope since the argument against providing an Emergency Egress System, capable of jettisoning the crew cabin from the payload, is that such a system would weigh too much. Instead, Shuttle crews must rely on an Emergency Egress System consisting of a telescopic pole attached to the mid-deck bulkhead, which can only be used at sub-Mach speeds, thereby severely restricting the availability of crew abort options. The current Shuttle configuration was the basis for many of the recommendations made by the Columbia Accident Investigation Board (CAIB) [2], which recommended the crew be separated from the payload. This configuration is achieved in the JLS architecture by carrying the payload separately, below the crew vehicle.

Ares I configuration vs. Jupiter System

It may appear the Ares I configuration achieves a high safety margin by producing a "crew only" (Ares I) and a "cargo only" (Ares V) launch system. However, in reality the team tasked with designing Orion has been forced to remove safety features and lunar capabilities to achieve a weight within the launch capabilities of Ares I. Such a program of weight reduction poses a significant danger to crews, given mission risk increases exponentially with distance from the Earth. It is important to remember much of the hardware developed now will be used for future missions to Mars. Any reductions in safety margins therefore will increase the LOC numbers significantly. The JLS vehicles, however, by virtue of their capability of placing additional mass in LEO, are able to reduce the probability of a LOC event to less than 1 in 1,000 (compared with the demonstrated 1 in 87 LOC rate for the Shuttle). Based on four launches a year, a 1 in 1,000 LOC event would occur once every 250 years. A counter-argument NASA might make to the DIRECT approach is that Ares I is rated to an LOC event once every 350 years, but the DIRECT team would fire right back and argue the cost of achieving such a safety margin is offset by the added safety factor of using a common launch system. Such an approach obviously results in a higher flight rate, which in turn provides additional knowledge of the system, a priceless factor in implementing risk mitigation systems. In fact, the DIRECT team might question NASA's quoted "once in every 350 years" LOC event by pointing to the safety features provided by the Jupiter 120 abort capabilities, which exceed the safe crew abort g-limits of Ares I.

Transitioning from Shuttle to Jupiter

The DIRECT transition from the Shuttle to the first Jupiter 120 launch is depicted in Figure 8.6. It is envisaged the first Jupiter 120 missions would be to the ISS. Since the DIRECT approach will reuse existing Space Shuttle infrastructure, the first crewed Jupiter 120 flights could commence in 2012, a full three years ahead of the first launch of Ares I. The first Jupiter 120 ISS flights would serve to replace ISS core components and to break in the new LV but, given Jupiter 120's high mission mass capability, it is also possible it could be utilized for unmanned missions such as launching sample return vehicles to Mars.

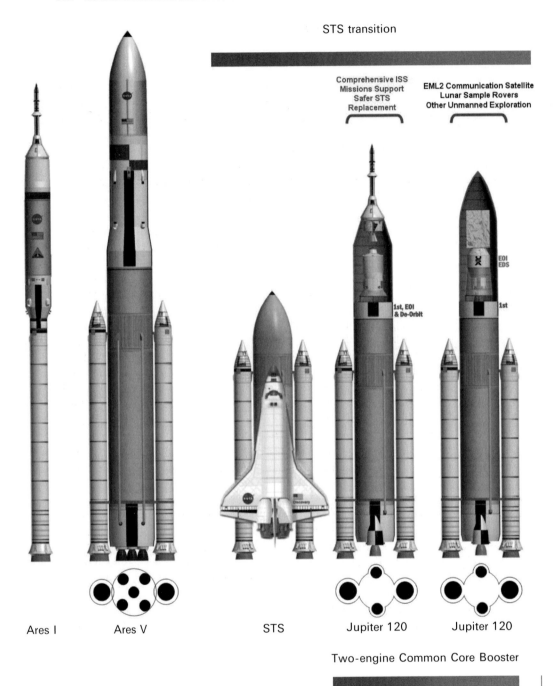

Figure 8.6. DIRECT transition from the Space Shuttle to the Jupiter launch system.

DIRECT'S LUNAR ARCHITECTURE

ESAS architecture concerns

Before describing the DIRECT lunar architecture, it is useful to review the ESAS recommended architecture which involves staging the mission in Earth orbit, followed by a LOR of the LSAM while Orion remains in lunar orbit for the return to Earth.

One of the problems of the ESAS architecture is the use of Earth orbit as a staging point, resulting in few available launch windows. The reason for this restriction of launch windows can be explained by intricate orbital mechanics, but since the complexity of such an explanation is beyond the scope of this publication, a layman's account is provided here.

To place a spacecraft into a $28°$ Earth inclination parking orbit, a due east launch from Kennedy Space Center (KSC) is used, but the equatorial bulge of the Earth results in the orbital plane rotating westward between $5°$ to $7°$ per day. The Moon, on the other hand, moves eastward in its orbit around the Earth by about $12°$ per day and the result of these orbital movements is only three Earth-to-Moon launch windows each month. This number may be reduced even further if a specific Sun orientation is required for a specific lunar location.

The Sun orientation factor becomes extremely important to mission planners due to the solar cycle on the lunar surface. While from the Earth we see one side of the Moon in sunlight, the Moon actually encounters sunlight on most of its surface over a 29-day interval, while it orbits the Earth. One solar cycle, therefore, which lasts 24 hours on Earth, takes approximately 29 days on the Moon. This causes big differences in the absence or presence of solar radiation, which in turn causes changes in the thermal environment the spacecraft must be subjected to. It was for this reason the Apollo missions landed on the Moon shortly after lunar dawn and departed shortly before lunar afternoon, thereby ensuring the spacecraft operated within a safe thermal range.

The problem of relying on so few launch windows is KSC weather can often cause a launch to be waved off. Using the ESAS architecture means mission elements already in orbit must continue to orbit for several weeks or more until weather permits a launch. Compounding the issue is the 14-day on-orbit lifespan of the EDS which, in the event of a launch window closing, may lose too much propellant as a result of boil-off prior to the next launch opportunity. As a result of so few launch windows, the DIRECT team and many space analysts argue that by utilizing the present ESAS architecture NASA may find itself under pressure to launch, a factor mentioned as a contributing cause in both the Challenger [1] and Columbia investigations [2].

Global access anytime return

One of NASA's stated objectives for the VSE [4] was to deliver a *global access anytime return* architecture. However, the global access component was removed by NASA management shortly after the limitations of mass requirements and the Ares I/Ares V EOR–LOR profile were realized and NASA decided instead to restrict its exploration objectives to a polar location [6].

DIRECT architecture advantages

The DIRECT Advanced Upper Stage (AUS) will use a pair of Pratt & Whitney Rocketdyne J-2X vacuum-optimized engines for the ascent and Earth departure phases of the mission. In the DIRECT the AUS will be launched first and may have to orbit for extended durations until the crew rendezvous, an architecture design requiring the use of an Integrated Cryogenic Evolved Stage (ICES) EDS, combined with an active cooling system to reduce boil-off to 0.01% per day. The AUS, once the crew have completed the rendezvous, will serve as the EDS and represents the stage enabling the DIRECT architecture to match the ESAS design.

The DIRECT fulfills the ESAS recommendations of utilizing two launches to place all lunar mission components in orbit but, whereas the ESAS requires one small Ares I for crew and one large Ares V for cargo, the DIRECT uses one reconfigurable vehicle, an approach solving many development problems. For example, building two different LVs requires two separate development programs, two operations support programs, and two manufacturing infrastructures, but with the DIRECT approach, utilizing existing Shuttle hardware and infrastructure, there are no such requirements, thereby saving money and time.

Mission timetable

Because DIRECT utilizes so much existing hardware and requires so little development of new systems, it will be able to commence the first crewed flights of Orion as early as 2012, and manned lunar exploration as soon as 2017, more than three years ahead of NASA's current architecture. This timeline significantly closes the gap between the retirement of the Space Shuttle, planned for 2010, and NASA's first manned flight of Orion which has now slipped to at least September 2015 and may slip still further to April 2016. However, this is not the whole story as it is likely new technologies will have to be developed and flight-ratified to satisfy NASA's architecture. This will result in the first Ares I/Orion flight being even further delayed.

ADVANTAGES AND DISADVANTAGES OF DIRECT

Apart from bringing forward the first manned Orion flight by three years, the primary advantage of the DIRECT is the increased safety and performance margin of the Jupiter 120 over Ares I. This is primarily by virtue of the dual main engine configuration of the Jupiter LV, providing Orion's crew with the capability to survive an engine-out scenario during ascent to orbit. In contrast, a main engine-out scenario on Ares I would most likely result in a LOM, requiring a potentially dangerous abort maneuver. Furthermore, although both the Jupiter LV and Ares I have a Launch

Abort System (LAS), the Jupiter LV LAS uses flight-proven systems at lower load levels, thereby increasing crew safety.

Another advantage is the flexibility inherent in the DIRECT architecture, offering a number of options not available to the ALS. For example, the Jupiter 2 series of vehicles have the capability of sending Orion on an Earth departure trajectory in one direct ascent launch, thereby offering mission planners architectural alternatives and mission staging points unavailable to the Ares I/Orion configuration.

Such architectural flexibility also offers mission planners the option of stationing a reusable and refuelable LSAM at Earth–Moon Lagrange Point 1 (EML1) for rendezvous with Orion. If a propellant depot were to be stationed at EML1, it would be possible to rotate a lunar surface crew using just one direct ascent launch of a Jupiter 221 containing just the crew in Orion. Using this architecture, all Orion would need to do would be to attain EML1 orbit and dock with the LSAM, then, after returning from the lunar surface, the crew would transfer to Orion and return to Earth. Another advantage of such a cost-effective architecture is that it would provide a daily departure window from the Earth and an uninterrupted return window from any location on the surface of the Moon.

An obvious advantage of the DIRECT over Constellation is the elimination of cost associated with developing two separate launch systems and infrastructures. The money saved by using the existing STS workforce and infrastructure can instead be applied to accelerating the development of the lunar phase components of the VSE, such as the LSAM.

The DIRECT program, if implemented, would not be immune from the usual problems associated with a complex development program. However, its proponents argue the robustness of a program that significantly reduces up-front development costs by re-sequencing available mission elements would enable results to be realized sooner and with less risk than the ESAS plan.

POLITICS

Despite the benefits of DIRECT, it is likely Constellation will continue to move ahead, although perhaps not in its original configuration. Under the existing schedule, despite a new presidency in 2009, most of the current members of Congress who accepted the Ares I/Ares V concept will still be there when Ares I becomes operational, and Ares V development begins. Since Congress will probably feel a sense of obligation to finish what it had started, it is probable Ares V will be built. However, because of the redesign of Ares I and because the development start of Ares V has moved so far to the right, it is now unlikely those Congress members who voted for the Ares I/Ares V concept will still be in their seats once development *does* begin. Instead, the Ares I/Ares V Congress members will have been replaced by those who have no commitment to either the VSE or the Ares I/Ares V.

Because of the slip in development, Ares V will not become operational until three or more Presidential administrations in the future, at which juncture the interest of Congress will be focused upon ensuring Ares I and Orion become operational sometime in 2015 or 2016. Remember, by this time the U.S. may have been paying the Russians through the nose to send NASA astronauts to the ISS. It is possible this scenario could be avoided if Commercial Orbital Transportation System (COTS) contracts, such as the one with SpaceX, come to fruition. In either case, the most likely scenario is Congress will fund only one LV, which means if we want to go to the Moon and Mars, the LV has to be capable of transporting humans beyond LEO. It was essentially this political background that spurred Tierney's team to propose DIRECT.

RESPONSE TO DIRECT

Unsurprisingly, publication of the DIRECT architecture resulted in vigorous discussions among space enthusiasts, many of whom openly wondered why NASA would not adopt what appeared a much more efficient and quicker means of returning to the Moon.

> "Good ideas, in the long run, sell themselves."
>
> Michael Griffin, NASA Administrator, *Heinlein Centennial*, July 16th, 2005

Some may take Griffin's quote and apply it to the DIRECT approach. However, Griffin, in the face of so much discussion directed at the feasibility of Constellation, felt compelled to defend it.

On January 22nd, 2008 Griffin delivered a speech to the Space Transportation Association addressing the concerns regarding the operations, risks, reliability, and costs of the Constellation architecture. Griffin began by reminding his audience the goals of the Constellation architecture were guided by President Bush's VSE speech [4, 5]. Griffin then reiterated NASA's goal of returning Americans to the Moon no later than 2020, to launch Orion as close to 2010 as possible, and enabling humans to land on Mars, acknowledging no policy is perfect and no plan would please everyone. Before addressing the choice of space architecture, Griffin restated NASA's exploration policy and its implications for returning to the Moon and Mars. He emphasized the intent of the Constellation architecture was not simply to design a system to replace the Shuttle, but to deliver a LV capable of carrying astronauts to the Moon and beyond. Griffin then continued to remind his audience of the commitment to the ISS, the longevity of launch systems, and the problems with policy and legislation, before finally tackling the meat and bones of the choice of architecture, which he began by stating:

> "However, this architectural approach carries significant liabilities when we consider the broader requirements of the policy framework discussed earlier. As with the single-launch architecture, dual-launch EOR of identical vehicles is vastly over-designed for ISS logistics. It is one thing to design a lunar transportation system and, if necessary, use it to service ISS while accepting some

reduction in cost-effectiveness relative to a system optimized for LEO access. As noted earlier, such a plan backstops the requirement to sustain ISS without offering government competition in what we hope will prove to be a commercial market niche. But it is quite another thing to render government logistics support to ISS so expensive that the Station is immediately judged to be not worth the cost of its support. Dual-launch EOR with vehicles of similar payload class does not meet the requirement to support the ISS in any sort of cost-effective manner."

To the disappointment of the DIRECT team, Griffin seemed to have missed the point on the subject of *cost-effectiveness* or chosen not to address it. According to the DIRECT publication, the Jupiter 120 is a superior LV to Ares I in terms of life cycle cost and implementation time even if the engineering flaws of NASA's LV are ignored. Another cost advantage Jupiter 120 has over Ares I is its capability of launching an ISS logistics module, along with a crew, to the Commercial Orbital Transportation Services (COTS)-servicing timeframe.

Another aspect Griffin chose to overlook was the issue of the COTS timeline. Since the DIRECT plan would provide the services of the Jupiter 120 at an earlier date than Ares I, it would be possible to leave the ISS-servicing role entirely to COTS. This would permit the maturing of key spacecraft systems such as the addition of the US required for the Jupiter 232, which, thanks to its single-launch capability, would permit more cost-effective exploration architectures.

Griffin then defended the Constellation architecture, stating:

"So if we want a lunar architecture that looks back to the ISS LEO logistics requirement, and forward to the first Mars missions, it becomes apparent the best approach is a dual-launch EOR mission, but with the total payload split unequally. The smaller launch vehicle puts a crew in LEO every time it flies, whether they are going to the ISS or to the Moon. The larger launch vehicle puts the lunar cargo in orbit. After rendezvous and docking they are off to their final destination."

Again, Griffin seemed to have evaded the matter of logistics issues, since, when considering an expansive lunar exploration program, the obvious strategy is to place all the propellant in orbit first, followed by the crew. This approach was conceived by the father of the Apollo program, Wernher von Braun, who described the architecture to Deputy Director of the Office of Manned Space Flight Dr. Joseph Shea in June 1962:

"Let me point out again that we at the Marshall Space Flight Center consider the Earth Orbit Rendezvous Mode entirely feasible. Specifically, we found the Tanking Mode substantially superior to the Connecting Mode. Compared to the Lunar Orbit Rendezvous Mode, it seems to offer a somewhat greater performance margin."

PURSUING CONSTELLATION

Another factor implicated in pursing the VSE is the question of legacy. It is probable that 2008 will be Griffin's final year as NASA administrator. Clearly, one of the biggest impacts Griffin has had on NASA, and the future of the American human spaceflight program, was his implementation of Constellation. The architecture which he approved has been discussed, debated, and precipitated the suggestion of alternate architectures, such as the one described in this chapter.

However, on many levels Constellation achieves the objectives of the VSE and ESAS [4, 6]. Although it might be highly desirable from an engineering perspective to have a multi-role architecture, in reality the NASA budget cannot support multiple architectures. Equally, given how far the design and development of Orion and Ares I has progressed to date, it is simply not financially possible to change now and adopt another architecture.

For all the criticism leveled against it, Constellation represents an evolution of the manned space program that looks beyond the Space Shuttle and the ISS. It also presents a logical plan for achieving completion of the ISS and facilitating a return to the Moon, thereby enabling humans to prepare for manned missions to Mars and beyond. Critics point to Constellation not meeting its performance levels and exceeding expenditure, but spaceflight is a government activity and there is hardly any government activity operating within desired performance levels, much less meeting desired expenditures.

Despite the criticisms, Constellation represents the best architecture that meets all of the requirements of space policy, budget restrictions, and engineering constraining NASA in the post-Shuttle era. It does not satisfy everyone, but Constellation enables achievement of the VSE goals. Furthermore, Constellation enables the United States to maintain its role as a world leader in the realm of manned spaceflight. By implementing ESAS, Griffin has perhaps ensured his legacy, namely the fear that by not pursing the VSE the United States will become irrelevant.

REFERENCES

[1]   *Report of the Presidential Commission of the Space Shuttle Challenger Accident* (June 6, 1986).
[2]   NASA. *Columbia Accident Investigation Report* (August 26, 2003). NASA, Washington, D.C. Available at *http://caib.nasa. gov/news/report/default.html*
[3]   DIRECT website at *http://www.directlauncher.com/*
[4]   President George W. Bush, *The Vision for Space Exploration* (January 14, 2004). NASA, Washington, D.C. Available at *http://www.nasa.gov/mission_ pages/exploration/main/ index.html*
[5]   NASA. *President's Commission on Implementation of United States Exploration Policy* (June 2004). NASA, Washington, D.C. Available at *http://www.nasa.gov/pdf/60736main_ M2_report_small.pdf*

[6]   NASA. *Exploration Systems Architecture Study: Final Report*, NASA-TM-2005-214062. NASA, Washington, D.C. (November 2005). Available at *http://www.nasa.gov/mission pages/exploration/news/ESASreport.html*

9

Lunar exploration objectives

In the long run, a single-planet species will not survive. One day, I don't know when,
but one day, there will be more humans living off the Earth than on it.

NASA Administrator, Dr. Michael Griffin,
quoted in "Mars or Bust," *Rolling Stone*, 2006

Constellation Mission Statement for Lunar Exploration

Conduct robotic and human lunar expeditions to further science and to test new exploration approaches, technologies, and systems to enable future human exploration of Mars and other destinations.

Chapter 1 briefly described the reasons for returning to the Moon. The reasons, in mission planners' parlance referred to as *drivers*, include science, technology, exploration, and exploitation. For each driver there are a number of mission objectives, many of which were identified by the committee of the Exploration Systems Mission Directorate (ESMD) Strategic Roadmap Meeting (SRM) held at NASA Headquarters in February 2005.

The SRM committee comprised NASA Associate Administrators William Readdy and Craig Steidle, and members that included NASA astronaut Don Pettit. Starting with the President's Vision for Space Exploration (VSE) as a foundation, the committee defined lunar exploration objectives by asking what NASA and the United States intended to do on the Moon. Based on the language in the VSE, the SRM committee agreed the fundamental goal is to advance the United States' scientific security and economic interest through a robust space exploration program. As a result of this goal, the SRM identified the two primary lunar exploration objectives as performing science and developing new approaches to support sustained human

Table 9.1. Lunar exploration objectives.

Approaches to support sustained human exploration to Mars and other destinations	
Study long-duration human physiology	Develop, demonstrate, and innovate future Mars system and subsystems
Develop and utilize lunar resources and understand their mapping to Mars	Demonstrate increased maintainability, reliability, and supportability
Enable business opportunities	Perform technology test and verification
Enhance U.S. strategic interests	Demonstrate operational techniques
Science approaches	
Investigate the origin and evolution of the Moon	Utilize the unique features of the Moon as a platform
Perform astrobiology	Perform human health and fundamental biological science
Study the Moon as a guide to other planets	Perform resource-related science

exploration to Mars and other destinations. To achieve these objectives, a number of broad approaches were identified (Table 9.1).

Some of these approaches were later refined as a result of a broad-based effort in 2006, led by NASA, to characterize lunar exploration themes and objectives. These themes and objectives are referred to as the Global Exploration Strategy for the Moon. NASA used them as a starting point to create its lunar architecture, in turn leading to the definition of the reference missions described in Chapter 1, and the mission modes described in Chapter 7. The Global Exploration Strategy comprises six themes of lunar exploration:

1. Exploration preparation. To use the Moon to prepare for future human and robotic missions to Mars and other destinations.
2. Scientific knowledge. To pursue scientific activities addressing fundamental questions about the Earth, the solar system, the universe, and our place in them.
3. Sustained presence. To extend human presence to the Moon.
4. Economic expansion. To expand the Earth's economic sphere to encompass the Moon, and to pursue lunar activities with direct benefits to life on Earth.
5. Global partnership. To strengthen existing international partnerships and create new ones.
6. Inspiration. To engage, inspire, and educate the public.

Within the six themes there are 180 possible objectives, divided into 23 categories. It is beyond the scope of this book to describe the details of each theme. Furthermore, some themes, like economic expansion, are the subject of publications such as Harrison Schmitt's *Return to the Moon*, which describes the entrepreneurial future of the Moon [12]. The goal of this chapter, therefore, is to describe the surface activities that must be performed to meet the objectives of the themes most pertinent to the early outpost development, namely exploration preparation, scientific knowledge, and sustained presence.

The objectives of the aforementioned themes will be achieved through implementation of a multiple-mission architecture, conceived by the Lunar Architecture Team (LAT). For example, the first stage in meeting the objectives of exploration preparation is the Lunar Precursor Robotic Program (LPRP), described briefly in Chapter 7. Once the first LPRP stages have been completed, it will be the turn of humans to continue preparations to explore the lunar surface, achieved by conducting extended extravehicular activities (EVAs). During the course of their exploration preparations, astronauts will also conduct scientific activities and eventually begin the process of enabling a sustained human presence. Each of these stages is described in this chapter.

LUNAR PRECURSOR AND ROBOTIC PROGRAM

The LPRP will initially target locations being considered for human operations. Once this objective has been met, the robotic and human exploration effort will be integrated, commencing with astronauts establishing a habitat on the lunar surface. As the habitat develops into a permanent outpost, astronauts will conduct surface operations devoted to utilizing lunar resources, such as oxygen, to sustain humans living on the Moon. Other missions will be directed at recovering resources mined on the Moon, such as He-3, for use on Earth.

The broader science and exploration objectives of the LAT and those identified by the Lunar Exploration Science Working Group (LExSWG) [6] require global access to the lunar surface and the capability to explore the lunar surface with a mobile science platform. To achieve these objectives, the Lunar Reconnaissance Lander (LRL) and the Lunar Surface Explorer (LSE) have been developed.

Powered by a radioisotope generator, the LRL has the capability to land and operate anywhere on the lunar surface. It is also capable of accommodating diverse deployable science payloads, such as a rover.

The LSE is a tracked rover capable of supporting a variety of payload options. Together, the LRL and LSE can collect and cache samples, prepare samples for robotic return and survey potential sampling locations in advance of astronaut sorties. Furthermore, the LSE is capable of being deployed to locations inaccessible to human sorties, thereby enhancing the geographic coverage of the exploration program. Given the versatility of the LRL and LSE, it is envisaged astronauts will assign mechanical and analytical tasks to the lander and the rover, thereby permitting crewmembers more time to perform other exploration activities. For example, the lander and rover may be tasked to conduct surface operations in the shadowed floor of a polar crater, too cold for human sorties.

EXTRAVEHICULAR ACTIVITIES

The core of the VSE is exploration. To achieve exploration goals, astronauts will search for resources, learn how to work safely in a harsh environment, and explore the lunar surface. These activities (Table 9.2) will require sustained periods of EVA outside the protective environment of the outpost. EVAs will require astronauts to wear an Extravehicular Mobility Suit (EMU), comprising a space suit assembly and a Portable Life Support System (PLSS).

Table 9.2. EVA tasks during lunar sortie and outpost missions [2].

EVA	Description	EVA	Description
Site preparation	• Survey and stake-out • Rock removal • Smoothing • Clearing dust control areas • Establish navigation aids	Habitat installation	• Hole excavation • Transport hab modules to base site • Unload, locate, level • Deploy and inflate back filling
Shielding installation	• Support for regolith • Regolith bagging for radiation protection • Bag stacking and lifting • Clearing access paths	Science	• Sample collection • Installation of experiments • Location of mapping for geological survey • Establish observatories
Power Systems and Thermal Control System	• Site preparation • Unload equipment from lander and transport • Deploy and assemble • Radiator deployment and activation • Connect to distribution • Shielding excavation	Logistics	• Unload from hangar • Unpack • Transport • Transfer • Storage • Waste disposal • Storage of spent recyclables
Resource operations	• Resource process site set up (pressure vessel, plumbing, gas holding tanks, pumps, heat exchangers)	Lander operations	• Servicing/minor repairs • Refueling • Pre-launch and checkout • Relocation of fueling depot
Mining operations	• Equipment transportation to site • System set-up (bucket wheel excavator, conveyors, regolith-bagging equipment, sorters, separators • System relocation	Upkeep	• Inspection • Field checks and measurements • Replacement of systems/ subsystems • Repair of equipment

Apollo extravehicular activity suits

The EMUs (Figure 9.1) used by Apollo astronauts were subjected to significant wear and tear as a result of lunar dust. The dust caused the EMU's wrist rings, neck rings, zippers, hose connectors, and gloves to become abraded and stiff, and degraded the reflective surface of the helmet visor, despite the best efforts of the Apollo astronauts to keep their suits clean and lubricated. Another problem experienced by the Apollo crews was the effect of working against the suits' internal pressure. This issue caused all Apollo astronauts to suffer blisters, fingernail damage, and pronounced soreness of their hands as a result of having to grip tools while squeezing their hands against the suit pressure [7].

Constellation Space Suit System

The hostile environment of the lunar surface, combined with the challenges posed by the conditions of lunar gravity and the duration of surface excursions, will require a tremendously versatile and rugged suit (Table 9.3). Unsurprisingly, given the problems encountered by Apollo astronauts, the new class of astronauts will be hoping for

Figure 9.1. The extravehicular activity suits worn by Apollo astronauts. Image courtesy: NASA.

Table 9.3. Design features and technological requirements of lunar EMUs [2].

Technological requirements	Design features
Durable, high-strength, low-mass materials	Low-mass hybrid suit structure
Low-mass thermal storage system	Low mass, regenerative non-vent, thermal control
Long-life thermal micrometeoroid garment materials; over-garment impenetrable to dust	Full lunar dust-thermal-micrometeoroid cover layer protection system
Suitable bearing seal systems	Heavy-duty dust seals for suit bearings
Low-torque, low-maintenance, lower-torso joints	Lower-torso mobility systems for traversing lunar terrain
Durability boot sole materials	Boots designed for traversing lunar terrain
Variable transmittance electro-chromic systems	Integrated automatically adjusting sun visor

a more robust and dust-resistant suit. To that end, engineers are developing the Constellation Space Suit System (CSSS).

The hostile environment of the lunar surface, combined with the challenges posed by the conditions of lunar gravity and the duration of surface excursions, will require a tremendously versatile and rugged suit. Unsurprisingly, given the problems encountered by Apollo astronauts, the new class of Moon-walkers will be hoping for a more robust and dust-resistant system. To that end, engineers are developing the Constellation Space Suit System (CSSS), which will be capable of supporting a significant number of surface EVAs during potential six-month outpost expeditions. In fact, the Constellation Program will require two spacesuit system configurations to meet the requirements of Orion missions to the ISS and to the Moon. Configuration One will support dynamic events such as launch and landing while Configuration Two, which is described here, will support lunar surface activities.

Suit design considerations

Not only must the new suit provide crewmembers with a comfortable interior pressure and microclimate catering to a wide spectrum of surface activities, it must also provide sufficient mobility and flexibility to ensure maximum productivity. Other factors space suit designers must consider include reducing the risk of decompression. This means choosing a suit pressure low enough to ensure maximum flexibility, but not so low that a high-pressure differential exists between the lunar base pressure and suit pressure. Space suit designers must also ensure easy and rapid donning and doffing of the suit and provide crewmembers with a suit integrity protecting them from micrometeoroid penetration or puncture from sharp edges.

Mark III and REI space suits

NASA has been testing space suits since the mid-1990s, laying the foundation for the development of a lunar surface suit enabling astronauts to fulfill exploration tasks and scientific fieldwork. The result of space suit testing and development is the rear entry Mark III Hybrid Suit (H-Suit), consisting of a hard upper torso and soft material for the arms and legs (Figure 9.2, see color section). The suit also incorporates rotational and flexural bearings to enhance suit mobility at critical joints, such as the shoulders, wrists, waist, hips, and ankles. A further improvement on the Apollo suit is a pair of improved gloves derived from the Space Shuttle program, which increases hand dexterity and mobility while reducing hand fatigue (Figure 9.3, see color section).

Other improvements include the provision of an in-suit PLSS recharge capability and the integration of an Electronic Information System that may eventually lead to a heads-up display (HUD) capability, similar to the one used in fighter aircraft.

Although the H-Suit is a proven method, the system has several disadvantages. Not only is it bulky and heavy, thereby imposing a high metabolic requirement on its user, the suit also requires time-consuming pre-breathing and depressurization. An alternative to the H-Suit may be a suit developed by ILC Dover and the Advanced EVA team at NASA Johnson Space Center (JSC).

The Rear Entry I-Suit (REI), similar to the H-Suit, is a soft suit incorporating a limited number of bearings at the wrist, shoulder, upper arm, upper hip, and upper leg joints. It was designed to provide the lightest possible rear entry system and improved donning and doffing compared with a waist entry suit.

Testing the H-Suit and REI Suits

NASA, in conjunction with ILC Dover, has conducted extensive testing on the H-Suit and REI Suits during the Desert Research and Technology Study (RATS) remote field tests at Meteor Crater, Arizona. The RATS testing program evaluated science procedures, efficacy of the Liquid Cooling Garment (LCG), head-mounted display (HMD) capabilities, and global positioning systems (GPS). The science procedures, for example, included activities such as core sampling, hammering on a rock and collecting samples, providing evaluators with a better understanding of how the suits might function during various lunar field activities. Similar studies have been conducted at Johnson Space Center's Lunar Crater (Figure 9.4, see color section)

Bio-Suit

The H-Suit and REI Suit represent a significant improvement on the Apollo A7LB suit and will probably be worn by astronauts during the first missions to the Moon. However, astronauts embarking upon longer duration missions later in the outpost's development may have the opportunity to wear a suit set to revolutionize human space exploration.

Based on the concept of biomechanically and cybernetically augmenting human performance capacity, the Bio-Suit System (Figure 9.5, see color section), conceived by Professor Dava Newman of the Massachusetts Institute of Technology (MIT), is envisioned to function as a second skin by providing mechanical counter-pressure (MCP).

Whereas the lightest H-Suit or REI Suit will weigh at least 40 kilograms, Newman's skin-tight, alternative Bio-Suit will weigh as much as two or three orders of magnitude less. Constructed of spandex and nylon, the multi-layered suit hugs the body's contours like a second layer of skin and its MCP technology ensures constant pressure is applied to the surface of the body. This pressure is needed not only to counteract the vacuum of the lunar surface and maintain the body's homeostasis, but also to avoid blood pooling.

Maintaining an even pressure over the surface of the human body has, until now, been achieved by utilizing the bulky gas pressurization systems embodied by the H-Suit/REI Suit. Thanks to new MCP technology that works along lines of non-extension, or those lines along the body undergoing little stretching as the body moves, the Gas Pressurization System of the H-Suit is no longer necessary.

Another advantage of the Bio-Suit over the H-Suit is suit compromise. If the H-Suit suffers a puncture, the crewmember must return to the outpost and undergo decompression, whereas a small puncture of the skin of the Bio-Suit requires only a bandage.

The Bio-Suit will also help astronauts stave off the debilitating effects of osteoporosis and general deconditioning, thanks to its ability to provide crewmembers with resistance levels to maintain muscle and bone integrity. Perhaps the most striking capability of the Bio-Suit, by virtue of its flexibility, is its ability to enhance human performance on the lunar surface and thereby empower exploration of the Moon.

Bio-Suit instrumentation

Integrated into the Bio-Suit is a Bioinstrumentation System modeled on the Operational Bioinstrumentation System (OBS) developed in the 1970s at the University of Denver for use during Space Shuttle missions. The original OBS consisted of a signal conditioner, an EVA cable, a sternal harness, and three electrodes, a suite of components that will be significantly upgraded in the Bio-Suit, which will feature biosensors integrated into the textiles. The biosensors will measure biomedical signals such as heart functions, oxygen consumption, and body temperature, while a suite of biochemical sensors will provide information concerning body fluids, and dosimeters will measure the local radiation environment.

An EVA system fitted with wearable sensors becomes increasingly important as surface excursions become longer, EVA locations become more isolated, and activities become increasingly complex. The combination of these factors has the potential to increase the distraction and fatigue of the crewmember. However, should an accident occur in an isolated location, or a crewmember becomes excessively tired, having access to the information provided by a wearable sensor system may improve chances of survival.

SCIENCE ON THE MOON

Astronomy and astrophysics

194,068 square kilometers on the far side of the Moon have been set aside for the purpose of establishing a future Moon base for observatories. One of these, the Icarus Lunar Observatory Base (ILOB), will be located at the point farthest away from Earth and provide astronomers with the perfect location to establish optical and radio telescopes, thanks to the absence of atmospheric interference and sunlight. In addition to requiring astronauts to assemble and emplace the ILOB, servicing missions will also be required periodically to clean the optics. Astronauts will also be required to remove the lunar dust, which, charged by the solar wind, attaches electrostatically to the components of the observatory.

Astrophysicists will also find the lunar environment an ideal location to perform interferometry and gravitational wave interferometry, since both are ideally suited to the harsh lunar conditions. By locating radio interferometry antenna arrays on the far side of the Moon, shielded from Earth's radio noise, astrophysicists will be able to observe low frequencies not observed on Earth and may be able to collect data about exotic phenomena such as pulsars, black holes, and planetary radio emissions. Again, astronauts may be charged with the assembly of antenna arrays and be required for maintenance activities to ensure components continue to function.

Other interferometry studies will probably be directed at observing the universe in ultraviolet (UV), optical and infrared wavelengths, enabling astrophysicists to better observe extra-solar planets. Once again, these studies will be enabled by a telescope on the far side of the Moon which, thanks to the extremely low temperature, is an ideal area for conducing IR observations. In addition to interferometry, astrophysicists will be interested in placing a series of widely spaced detectors on the lunar surface to detect and study gravitational waves formed by merging supermassive black holes and binary compact objects.

Of particular importance to those living on the Moon will be studies aimed at defining, characterizing, and predicting the lunar radiation environment. Since the Moon is outside the Earth's magnetosphere and lacks an atmosphere, it is subject to constant bombardment by energetic solar particles and cosmic rays. By installing large high-energy cosmic ray detector arrays on the lunar surface, astrophysicists will be able to measure radiation striking the surface and perhaps provide a method to forecast serious radiation events. Once again, the emplacement of such equipment will be the task of astronauts living on the surface.

The relatively low seismic noise of the Moon may make it an appropriate location to search for a special, and hitherto unobserved, type of nuclear matter known as strange quark matter (SQM). Although SQM exists only on the pages of textbooks, it is proposed such matter is the most stable form of matter in the universe and may have been produced by the Big Bang itself. Astrophysicists keen on detecting this unique type of matter will install a network of seismometers, since the neutron stars thought to produce SQM often leave behind seismic signatures. If such a signature were to be observed, it would constitute a discovery of profound importance to the field of astrophysics.

Heliophysics

Another means of defining the radiation environment is the study of low-frequency radio astronomy observations, which can inform astronomers of potential solar particle events (SPEs). SPEs have the potential to kill lunar-bound astronauts and destroy orbiting assets such as the Crew Exploration Vehicle (CEV). Due to the ionospheric effects of the Earth's atmosphere, it has not been possible to perform these observations at frequencies below 10 MHz. By performing these observations over a number of years it may be possible to develop a space weather and radiation forecast that will greatly increase the safety of those working on the Moon.

Similar studies will be conducted using photon and particle imaging of the Earth's ionosphere, thermosphere, mesosphere, and magnetosphere to provide measurements that will help scientists to better understand how space weather affects the regions where military and civilian space operations occur in orbit around the Earth.

Lunar heliophysics studies may also be instigated to analyze the Sun's role in climate change. To do this, heliophysicists will collect photometric and spectral observations of earthshine in the electromagnetic and charged particle spectrum, which will provide an index of the Earth's reflectance. After a month of these observations, it will be possible to calculate the Earth's total reflectance, also known as *Bond albedo*, used to determine the amount of sunlight reaching the Earth. By comparing measurements of Bond albedo and solar output over many months and years heliophysicists will gradually develop an understanding of the dynamics of the Sun and the Earth's response, which in turn will help them to predict future climate change.

Earth observation

The surface of the Moon is the ideal location for a remote-sensing platform since it affords global observation of the Earth. Synthetic Aperture Radar (SAR) will utilize this feature to create highly accurate terrestrial topography, altimetry, and vegetation charts.

The lunar surface also provides the perfect vantage point for simultaneous observations of the Earth–Sun system, helping scientists to understand the reaction of the Earth's atmosphere to solar activity, in turn helping to estimate the effect of long-term solar changes on climate. The advantage of viewing the whole-Earth disk also enables the collation of information concerning land surface mineralogy, land use, land change, and biomass utilization.

Observations of the whole-Earth disk may also prove useful for monitoring ocean color, sea surface temperature, and atmospheric studies related to large dust emission events such as sandstorms in the Sahara. Lunar-based Interferometric Synthetic Aperture Radar (InSAR) will help scientists understand the dynamic response of major ice masses to climate change, as well as permitting the monitoring of sea ice extent and concentration in the polar regions. The whole-Earth perspective will also permit multispectral thermal infrared observations and subsequent atmospheric monitoring of events such as volcanic eruptions and fires.

Geology

One of the first geological investigations will probably be collating data on Moon-quakes, providing lunar inhabitants with a means to determine which areas are safe and which are liable to seismic activity (Figure 9.6, see color section). In addition to revealing site hazards, seismic exploration will also help geologists define mantle characteristics, identify buried lava tubes, and provide an insight into the processes involved in planetary evolution and crustal genesis.

Other geological variables of interest will include the characterization of endogenous (from the lunar interior) and exogenous (delivered externally) processes, which resulted in the deposition of volatiles on the lunar surface. Such studies will lead to predictive models of distribution of volatiles that may assist in the procurement of materials for *in situ* resource utilization (ISRU). Similar studies will be directed at characterizing the general geology of the Moon by investigating materials present in regolith to locate and identify elements useful in the development of a lunar outpost.

For geologists interested in impact cratering, there will be plenty of opportunities to investigate and characterize the nature of the heavy bombardment that has occurred during the Moon's lifetime by studying the morphology of old and freshly formed craters and their ejecta distribution. These studies will provide insight into the surface geology of the inner planets and will bring about a means of calibrating the ages of surfaces of other planetary bodies, such as Mars. Similar studies will be conducted on meteorite impacts to define the timing and composition of impactors upon the lunar surface. By studying meteorite impacts on the Moon, geologists may gain a better understanding of the impact history of the solar system, possibly provide clues to the thermal and aqueous history of early Earth, and obtain data supporting or rejecting the existence of past lunar life.

Lunar regolith will be investigated not only for the purpose of defining ISRU sites but also to help understand the nature and history of solar emissions, galactic cosmic rays (GCRs), and dust from interstellar medium, all of which are preserved in lunar regolith. The value of these investigations lies in uncovering the history of the Sun, since buried regolith may have preserved snapshots of solar radiation events. These snapshots may present geologists with a series of geological time capsules, providing a record of radiation, solar wind, and cosmic ray fluxes, information that may be used to track changes in the Sun. These studies may also be used to support heliophysicists' investigations of the evolution of the Galaxy and the history of Earth.

The study of regolith will also provide information permitting geologists to better characterize the process of space weathering of planetary bodies without air or water.

These studies will require mapping regolith maturity to identify regions of ancient and new regolith, data that will be used not only to understand the weathering process but also provide a reference point for understanding the potential hazards of resource mining on the Moon.

Another aspect of geological study is the issue of providing curatorial facilities and technologies to ensure samples returned to terrestrial geologists are not contaminated while on the Moon or during transport to Earth. The capability to curate samples on the Moon will apply not only to geological but also to biological samples and the products of resource extraction experiments.

Materials science

One of the primary efforts of materials scientists will be to design mitigation strategies aimed at ensuring the robust performance of hardware, particularly during extended stays of six months or more.

To achieve this, scientists will first study the unique aspects of the lunar environment. These studies will include investigations of fractional gravity, ultra-high vacuum, radiation bombardment, thermal cycling, and perhaps the lunar inhabitants' biggest headache: lunar dust. Characterizing the cumulative effects of the lunar environment will not only permit engineers to make informed decisions regarding the materials best suited to withstand extended use on the Moon but also provide a useful reference point for mission planners designing Martian mission architectures.

The locations materials scientists will visit for their investigations will include the sites of the Apollo, Lunakhod, and Surveyor spacecraft. These spacecraft have been exposed to solar radiation, lunar dust, and micrometeorites for more than three decades and will provide invaluable information about the effects of long-duration environmental exposure on spacecraft materials.

Physiological adaptation

Every astronaut who travels to the Moon in the next several decades will serve as a lab rat for terrestrial physiologists tasked with understanding the effects of extended lunar stays and ensuring the health and safety of crewmembers.

Although the effects that fractional gravity and radiation bombardment have upon the body are somewhat understood, others, such as the combined effects of the exploration environment, will require further study. In addition to characterizing the physiological response to extended lunar stays it will be necessary to study human factors, such as whole-body coordination strategies, to help astronauts deal with the effects of prolonged isolation and communication lag on performance and mission coordination.

Later, as the lunar outpost becomes more developed, research will be conducted on animals to determine the likely effects of the lunar environment on the ability of humans to reproduce normally. The outcome of these investigations will be crucial in determining if and how humans will be able to develop permanent colonies on planets such as Mars and moons such as Titan and Europa. The outpost will also be an ideal site for medical specialists to develop and deploy telemedicine equipment, define medical practices, diagnoses, and treatment strategies.

IN SITU **RESOURCE UTILIZATION**

ISRU is a process involving any hardware or operation exploiting and utilizing either natural or discarded *in situ* resources to create products and services for robotic and human exploration. For example, on the Moon, *natural in situ* resources include regolith, minerals, volatiles, metals, water, sunlight, and thermal gradients, whereas *discarded in situ* materials include the descent stage of the LSAM, fuel tanks, and crew trash. To make the most of ISRU, there are a number of products and services that must be considered (Table 9.4).

One of the first ISRU missions the crew will be tasked with will be ensuring their own survival. This will require protecting themselves from the harsh radiation environment by using regolith to bury the habitat. Once the habitat becomes established and ISRU processes become more developed, the crew will begin to process lunar materials to produce water and oxygen, used as a contingency supply for the Environmental Control Life Support System (ECLSS) and as a consumable source for EVA missions.

At this phase of lunar outpost development, it is likely *in situ* propellant production will have commenced. This will provide an extra margin of safety to lunar-bound astronauts as the extra fuel will help eliminate propellant loss due to leakage or increased boil-off from the LSAM ascent stage. The extra propellant will also be

Table 9.4. ISRU products and services.[a]

Site preparation and outpost deployment/emplacement
• Site surveying and resource mapping in order to determine usable resources • *In situ* water production for crew radiation protection • Area clearing to minimize risk of payload delivery and emplacement
Mission consumable production
• Complete life support and EVA closure by means of using *in situ* water resources • *In situ* propellant production for robotic and manned vehicles • Regeneration of fuel cell power consumables
Outpost growth and self-sufficiency
• Fabrication of structures using *in situ* materials • Fabrication of solar array and antennae • Thermal energy storage using processed regolith

[a] Adapted from Simon, T.; Sacksteder, K. NASA *in situ* resource utilization (ISRU) development and incorporation plans. Presentation at *Technology Exchange Conference, Galveston, TX, November 2007.*

used to fuel manned and unmanned surface rovers, thereby extending the range of planned lunar sorties.

The next phase of ISRU will be characterized by a more efficient power architecture, enabled by increased oxygen and hydrogen regeneration, thereby increasing night-time power and mobile fuel cell power requirements.

ISRU systems and technologies

Under the direction of NASA's Exploration Technology Development Program (ETDP), the systems and technologies enabling ISRU are presently at various stages of development. This section describes some systems and technologies required by crews as they embark upon the many steps towards realizing the efficacy of ISRU on the lunar surface.

Lunar regolith extraction

Whether it concerns propellant production, oxygen generation, or the mining of construction material most ISRU processes will require extraction of lunar regolith. This process will require the use of quarrying equipment similar to that found in terrestrial mines.

To excavate and transport materials it will be necessary to first develop dust-tolerant mechanisms and to fully understand the characteristics of lunar regolith and its behavior in one-sixth g. Once these objectives have been accomplished the next step will be to construct excavation equipment such as regolith excavators and haulers and surface mobility platforms. These systems will be sustained by high-cycle life, high-power density power systems and most will be under autonomous control (Figure 9.7).

The goal of developing and constructing regolith excavators will be pursued in conjunction with developing excavation subsystems such as regolith thermal management equipment, water, carbon dioxide, and gas separation subsystems, in addition to navigation and control software to enable automation of the excavation plant.

Astronauts will play a critical role in the development of extraction systems due to the requirement to build confidence in ISRU as early as possible during the growth of the outpost. Ensuring maturity of these systems will require extensive ground testing by astronauts and the development of multiple generations of hardware and systems.

Extracting water ice

Analysis of lunar samples returned by the Apollo missions confirms lunar soil is rich in oxygen, silicon, iron, calcium, magnesium, and titanium, each of which may be extracted from the regolith by various processing techniques. In addition to the aforementioned elements another use of ISRU will be extracting water from the ice-rich lunar regolith found in the cold-trap areas near the lunar South Pole. Latest analysis of data from Lunar Prospector's neutron spectrometer indicates there may

Figure 9.7. The Autonomous Drilling Rover may be used by astronauts to test regolith samples prior to excavation operations. Image courtesy: NASA.

be as much as 300 million tonnes of water ice contained in the permanently shadowed cold traps near the poles [1].

To extract water ice it will be necessary to construct a regolith reduction reactor, which will work by simply heating the regolith using solar energy. As the regolith is heated up to about 100°C, using low-temperature processing, any gases or ices present will be released, separated, and further processed. Additional heating of the regolith to temperatures of up to 1,200°C, using medium-temperature processing, will result in the release of volatile gases such as carbon dioxide (CO_2) and methane. Further heating of the regolith to 1,630°C by carbothermal reduction of lunar silicates (high-temperature processing) will produce oxygen, a process described later.

Developing construction materials

One method suggested for developing construction materials from the lunar regolith is *sintering*. Sintering is a process in which soil is heated and maintained at temperatures below its melting point, thereby permitting viscous flow on the grain surfaces to form necks between particles, resulting in the particles bonding together. Studies investigating sintering of lunar soil [13] returned by the Apollo 14 mission demonstrated that the process results in a porous and brittle material, although the cohesiveness of the solid can be increased by applying higher temperatures and greater pressures.

Further research [9] investigated the effects of modifying the sintering process in an attempt to increase the structural integrity of lunar materials by conducting tests on soils returned by Apollo missions 11, 15, and 16. Stronger materials were achieved as a result of tweaking the sintering process but the method that resulted in a material with the greatest mechanical properties was melting and casting.

To increase the flexural and tensile strengths of the resultant material it was suggested fiber reinforcements produced from lunar resources be added. It might be thought, due to the Moon's reduced gravity level and the reduced loading conditions, it would not be necessary to reinforce construction materials. However, this is not true due to the requirement to shield from such environmental effects as radiation, extreme temperature changes, and meteorite impacts.

Mining helium-3

In 1985, student engineers at the University of Wisconsin discovered that a sample of lunar soil, taken by Apollo 17 astronaut-scientist Harrison Schmitt from the rim of the Moon's Camelot crater, contained significant quantities of a special type of helium known as helium-3 (He-3). The discovery intrigued the scientific community due to He-3's unique atomic structure, making it a perfect candidate fuel for nuclear fusion, a process capable of generating considerable amounts of electrical power. In addition to being extremely potent, He-3 is non-polluting and has virtually no radioactive by-product, making it an ideal candidate as a 21st-century fuel source. Unfortunately, hardly any He-3 exists on Earth but scientists estimate there may be a million tonnes of He-3 on the Moon, which would be enough to supply the world's energy requirements for several thousand years. In fact, the equivalent of a single Space Shuttle payload (approximately 25 tonnes) could supply the entire United States energy requirements for a year.

> "Helium 3 fusion energy may be the key to future space exploration and settlement."
>
> Gerald Kulcinski, Director of the Fusion Technology Institute (FTI),
> University of Wisconsin

He-3 is deposited in the powdery lunar soil when solar wind, composed of a stream of charged particles emitted by the Sun, strikes the Moon. This process has continued for billions of years, resulting in a potential cash crop of He-3 ready to be strip-mined from the lunar surface.

One method scientists use to determine the amount of He-3 on the Moon is estimating the probable reserves of implanted He-3. This is done by calculating the quantity of electroconductive minerals and the degree of radiation imperfection of a mineral crystal lattice, determined by the maturity of the lunar regolith. For example, highly electroconductive metallic minerals (such as *ilmenite* lunar basalt containing a high proportion of titanium) maintain their crystal structure under surface solar radiation, whereas nonelectroconductive rock minerals lose their crystal structure and become amorphous.

Based on this method and direct measurements of regolith volume and thickness, scientists have categorized the different types of lunar rocks based on the amount of ilmenite and can thereby predict the abundance of He-3. For example, lunar rocks, referred to as *mare basalts* by lunar geologists, were divided into three categories based on the amount of ilmenite contained. Using data from the Apollo and Clementine missions, scientists predicted, by geological inference, the abundance of high-category titanium basalt (5%–10%) was approximately 53,000 tonnes on the near side of the Moon and 61,000 tonnes for the total lunar reserves. Probable reserves for moderate-category titanium basalt (3%–5%) were calculated to be 109,000 tonnes on the near side. Lowest category titanium basalt (1%–3%) was calculated at 143,000 tonnes [5].

The He-3 concentrations in lunar soil are 13 parts per billion (ppb) although lunar scientists postulate levels as high as 20 ppb to 30 ppb may exist. Although such values seem extremely small, the projected value of He-3 is $14,000 per gram. This means just 100 kg of the substance would be worth $140 million. By comparison, the current value of pure gold at today's price is $15,500 per kilogram.

To extract commercially viable quantities of He-3, it will be necessary to process significant quantities of lunar rock and soil. For example, to obtain 100 kg of He-3 it will be necessary to dig a two-kilometer square patch of the lunar surface to a depth of three meters. Although this sounds like a lot of work to extract such a small amount of fuel, it is worth noting that 100 kg of He-3 would be sufficient to provide power to a city the size of Dallas for an entire year [14]!

A team of American scientists intends to oversee the He-3 extraction process and have already acquired the mineral rights for 95% of the Earth-facing side of the Moon, the polar regions, and 50% of the far side of the Moon. Former NASA astronaut Dr. Joseph Resnick and current NASA consultants Dr. Timothy R. O'Neill and Guy Cramer (the "ROC" team) recently announced they had secured more than 75% of the lunar mineral rights to allow for the extraction of He-3 for the advancement of space exploration and more efficient energy production. Having secured the mineral rights[1] the ROC team intend to ensure the Moon does not become scarred by strip-mining visible to orbiting space stations by factoring in environmental and visual preservation measures to preserve the lunar surface in the same state it was prior to mining.

Extracting He-3 will require teams of astronauts trained in the art of mining and the handling of robotic equipment. Two strategies for mining, processing, and refining the lunar regolith have been proposed by the University of Wisconsin.

One of these strategies is referred to as a *rectilinear strategy* [17], a process similar to the method used to mine deposits of relatively disaggregated materials on the surface of the Earth. On the Moon, this process would require mined regolith to be transported by bucket wheel into a mineral processor, in which larger fragments

---

[1] Space Law does not allow countries to have land ownership on planets or moons but it does allow for the mineral rights to be obtained by individuals and companies.

would be rejected by sequential sieving and the finer material would be taken to a volatile extraction unit. Once extracted, the volatiles would be placed into tanks and transported to a refining plant.

The alternative strategy, known as *spiral mining*, is a strange fusion of large-scale terrestrial open-pit mining and circular irrigation. Spiral mining requires the placement of a mineral processor unit attached to the end of a telescoping arm, which in turn is attached to a mobile central station. Using this process, the mining and processing would occur in an outward spiral with volatile materials being extracted from the regolith before being transported to the central station for refining. In this system, the central station would house all the refining, power, control, and habitation elements which would move together once the telescoping arm had reached its limit of operation, at which point the whole system would move to the next mining area.

Regardless of which process is used to mine regolith, the means of processing will probably be very similar. Studies investigating the design of a lunar regolith processor have been conducted at the University of Wisconsin's Fusion Technology Institute by Dr. Sviatoslavsky [16, 17]. Sviatoslavsky's Mark II Miner, with its arm-mounted bucket wheel excavator, is designed to "eat" its way through lunar regolith in a 3 meter deep and 11 meter wide trench. In the Mark II Miner's mode of operation, mined material flows out of the bucket wheel trays onto a conveyor belt and onto a series of progressively smaller sieves which filter out fine particles. The particles are then transported to the Mark II Miner's pressurized interior. Once inside, the sieved regolith is filtered again by a stream of gas separating fine particles into even finer particles, which are then fluidized, before being transported through a heat exchanger. After being exposed to heat for 20 seconds, the very fine regolith particles are pumped into high-pressure cylinders and transported to the refinery.

Once refined, the He-3, which would now be in the form of either pressurized gas or cryogenic liquid, would be ready to be shipped to Earth by unmanned spacecraft fueled by lunar-derived propellant systems.

OXYGEN PRODUCTION

To ensure the settling and evolution of a permanent and autonomous lunar outpost it will be necessary to produce lunar liquid oxygen (LLOX) utilizing indigenous resources. Not only will such *in situ* utilization result in cost savings on propellant for transportation systems, it will also support those living on the lunar surface. Fortunately, oxygen is the most abundant element on the Moon, but the problem is it must be extracted from lunar rock and regolith, where it exists in chemical combination with elements such as iron and titanium.

Factors to consider

The goal of this section is to describe some of the more promising concepts proposed for the production of LLOX, but before reviewing some of these it is useful to consider briefly the factors that must be considered.

Energy requirements are obviously important, so engineers must decide whether to use solar electric or nuclear electric power sources. Engineers must also consider the simplicity of the production process, the costs of importing consumable reactants, and whether the sites to be mined comprise rocks or soils. This latter consideration is important since different types of lunar ores require different processes to extract the oxygen. Furthermore, the different phases of production when applied to certain lunar ore types result in the liberation of potentially harmful substances. For example, *troilite*, when roasted in hydrogen or oxygen, readily decomposes and liberates sulfur.

Oxygen production processes

Reduction

One of the simplest processes to produce oxygen is to react lunar minerals, such as ilmenite, with gases. One way this may be achieved is to reduce ilmenite with hydrogen, a process first suggested by Williams, McKay, Giles, and Bunch in 1979 [19]. Williams and his team proposed reducing ilmenite by raising the reaction temperature to between 700°C and 1,000°C using hydrogen as a reductant. At this temperature, water would be produced, electrolyzed into hydrogen and oxygen, and the hydrogen would be recycled. In anticipation of this process being used in the future, Carbotek Inc., of Houston, Texas, has patented an ilmenite, hydrogen reduction technique using a reactor comprising three stacked fluidized beds.

Although the chemical process of hydrogen reduction of ilmenite is fairly simple and has been demonstrated in terrestrial laboratories, it is not known how practical the process will be in one-sixth terrestrial gravity. Other problems engineers must solve include eliminating the effects of ilmenite impurities, such as magnesium, upon the kinetics of ilmenite reduction with hydrogen. They must also develop a means to retain the hot hydrogen within the system, since even a small leak on the Moon could cause large quantities of gas to be exhausted due to the hard-vacuum conditions of the Moon.

Another gas engineers may use as a reductant to process ilmenite is CO_2, a process described by Zhao and Shadman [20] using a system similar to Carbotek's fluidized bed. One disadvantage of this system are the consequences of the resulting endothermic reaction which would consume considerable energy, although this shortcoming would be offset in part by the recovery of carbon from the ilmenite, possibly reducing consumption losses. Another consideration for the use of CO_2 as a reductant is that the rate of reduction is less than with hydrogen, although studies have demonstrated that for a given temperature, the difference is less than an order of magnitude.

A third reductant candidate for ilmenite reduction is methane, using a process similar to the one used for using hydrogen and CO_2 as reductants [4]. In this system methane is produced by reacting CO_2 and hydrogen products with extra quantities of hydrogen at temperatures between 800°C and 1,000°C, using nickel as a catalyst. The end-product of this series of chemical events is water, which is simply dissociated to produce oxygen and hydrogen.

Carbothermal reduction

Carbothermal reduction is a cutting edge oxygen production process combining some of the chemical processes of steel production with the electrolysis and thermolysis of water in which molten reactant is reduced using carbon in various forms [11]. Carbothermal reduction might be used to process simple oxide minerals such as ilmenite, although the very high temperatures required for the process may prove problematic.

Another approach may utilize devolatilized carbon to reduce ilmenite, which would require smelting the ilmenite, decarburizing iron, and reforming hydrocarbon. This process would ultimately produce water, which would then be electrolyzed to yield oxygen and recyclable hydrogen.

In common with many other oxygen production techniques, the problems inherent in carbothermal reduction are the high temperatures required, the resulting corrosive nature of the melt, and the subsequent wear and tear on the reactor linings. However, given the estimated 22 ppm concentration of indigenous solar wind–implanted carbon in lunar soil, the carbothermal process might yet prove to be worth pursuing.

Oxygen from glass

Another source of oxygen on the Moon is found in lunar glass, one of the most abundant constituents of the lunar regolith, especially in the mare regions. Most lunar glass originated as a consequence of melting produced by volcanic activity and meteorite impacts. These processes resulted in impact glass fusing together rock and mineral fragments into agglutinates which, in the mare regions, constitute more than half the lunar soil. It has been demonstrated [8] that by using hydrogen as a reductant it is possible to reduce lunar glass to iron and water, which can then be easily hydrolyzed or electrolyzed to yield oxygen and recyclable hydrogen. Once again, oxygen might be produced by utilizing the reduction process mentioned previously, although some studies have suggested some agglutinitic glass may be reduced using magnetic separation, due to its high titanium content [18].

Electrolysis and pyrochemical techniques

Oxygen may also be produced by electrolysis which, in its most basic form, simply entails placing two electrodes in a vat of molten silicate and sending a current between the electrodes. The result of this simple one-step process, requiring no moving parts or reagents, is oxygen derived at the anode and metal derived at the cathode. The efficiency of such a system is dependent upon the fraction of the current required to produce oxygen, which in turn depends on the concentration of iron in the silicate melt. Studies have shown that to produce oxygen efficiently, the silicate melt should be less than 5% FeO [5] and ideally have low concentrations of silica, aluminum, and FeO, since these elements increase the energy requirements.

Although the silica melt process sounds simple, it is a process limited in terms of industrial experience and due to the extremely high temperatures involved (between 1,300°C and 1,700°C), the problems of anode stability and corrosion are significant. To alleviate the problems of such high temperatures, using a flux to dissolve the silicate has been suggested, a process which would reduce operating temperatures and increase electrolyte conductance, thereby resulting in lower energy consumption. A similar process is already used on Earth in the manufacture of aluminum.

Pyrolysis

Pyrolysis requires the application of very high temperatures to induce chemical change and partial decomposition of metal oxides. The temperatures, usually in the range of 2,000°C and 10,000°C, are generated by systems such as plasma torches, electron beams, and solar furnaces.

Lower temperatures are used for vapor pyrolysis, a process that literally vaporizes the material, transforming it into oxygen-bearing compounds, after which the gas is rapidly cooled and everything except the oxygen is condensed into either a liquid or solid. One obvious advantage of pyrolysis is it requires no consumables from Earth since the process is entirely dependent on space resources, such as solar energy and the hard vacuum of the lunar atmosphere.

At slightly higher temperatures, between 3,000°C and 6,000°C, ilmenite can be completely dissociated into iron, titanium, and oxygen, although the process, which would require a solar furnace, requires considerable development.

At higher temperatures, usually between 7,000°C and 10,000°C, pyrolysis results in ionization. For example, at 7,700°C nearly all the metallic dissociation products, such as iron, titanium, and magnesium, are ionized and extracted from the vapor by electrostatic or electromagnetic fields, leaving the neutral oxygen to be collected [15].

Thermal recovery of oxygen

Another very simple process for producing oxygen is to roast lunar soil to temperatures of 900°C [3]. At this temperature all the hydrogen contained in the lunar soil is released and can react with oxide minerals to produce water in a process similar to that used to reduce ilmenite by hydrogen. Electrolysis may then be used to dissociate water into hydrogen and oxygen. Although the process is simple and requires no difficult development, a major disadvantage is the huge quantity of lunar soil that would require processing. However, this shortcoming may be slightly offset by the possibility of co-producing oxygen, helium, and hydrogen in the recovery process.

Fluorination

The process of fluorination liberates oxygen by using fluorine as a reagent to decompose silicates, thereby separating elemental oxygen from the lunar soil. On Earth, geochemists have successfully used fluorine gas to extract oxygen from terrestrial rock samples and from original and simulated lunar material [10]. Although these early studies yielded only small amounts of oxygen, subsequent studies investigating the effects of fluorination at different temperatures upon lunar soil simulants with different elemental properties demonstrated about 80% of oxygen could be extracted from lunar soil.

One of the problems of fluorination is the high temperature required and the corrosive effect fluorine has upon certain metals and metal alloys exposed to fluorine gas at varying temperatures and atmospheric pressures. In fact, the only non-metallic materials capable of resisting fluorine attack at high temperatures are highly sintered clay and poly-tetrafluorethylene (PFTE).

The processes described provide an overview of some of the candidate processes that may be considered to produce oxygen on the Moon. Some are relatively simple but may be difficult to implement on the surface of the Moon. Others are complex but are relatively untested, and yet others may not be economically viable.

ISRU DEVELOPMENT AND INTEGRATION STRATEGY

NASA intends to develop ISRU through a phased ground development, without requiring the LPRP missions. This development will occur in four phases, each lasting between two and three years. Phase I will demonstrate feasibility, Phase II will evolve systems developed with improved technologies, and Phase III will develop one or more systems to Technological Readiness Level 6 (TRL6) prior to the start of flight development. Finally, Phase IV will comprise flight development for outpost deployment.

The task of attaining the lunar exploration objectives described in this chapter will be a challenge. However, the phased approach envisioned by NASA will incrementally extend the lunar exploration campaign and gradually basic and intermediate science and exploration objectives will be met, laying the foundation for more sophisticated surface missions to be implemented. As more and more exploration objectives are achieved, full exploitation of the Moon's science and exploration potential will be realized. The experience gained from fulfilling the objectives will not only achieve a comprehensive scientific understanding of the Moon but also prepare humans for the exploration challenges for manned missions to Mars.

REFERENCES

[1] Binder, A. Update on Lunar Prospector results. Presentation at the *Space Resources Roundtable II, Colorado School of Mines, Golden, Colorado, November 8–10. 2000.*

[2] Bufkin, A.; Tri, T.O.; Trevino, R.C. EVA concerns for future lunar base. *Second Conference on Lunar Bases and Space Activities of the 21st Century*, Paper No. LBS-88-214. Lunar and Planetary Institute, Houston, TX (1988).

[3] Christiansen, E.L.; Euker, H.; Maples, K.; Simonds, C.H.; Zimprich, S.; Dowman, M.W.; and Stovall, M. *Conceptual Design of a Lunar Oxygen Pilot Plant*, EEI Rept. No. 88-182. Eagle Engineering, Houston, TX.

[4] Friedlander, H.N. Analysis of alternate hydrogen sources for lunar manufacture. In: W.W. Mendell (ed.), *Lunar Bases and Space Activities of the 21st Century*, pp. 611–618. Lunar and Planetary Institute, Houston, TX (1985).

[5] Haskin, L.A.; Colson, R.O.; Lindstron, D.J.; Lewis, R.H.; Semkow, K.W. Electrolytic smelting of lunar rock for oxygen, iron and silicon. In: W.W. Mendell (ed.), *Second Conference on Lunar Bases and Space Activities of the 21st Century*, NASA-CP-3166, Vol. 2, pp. 411–422 (1992).

[6] Kring, D.A. *et al. LExSWG: Lunar Surface Exploration Strategy*, Final Report. Lunar and Planetary Institute, Waltham, MA.

[7] Larson, W.; Pranke, L. (eds.). *Human Space Flight: Analysis and Design*, Space Technology Series, McGraw-Hill, New York (1999).

[8] McKay, D.S.; Morris, R.V.; Jurewecz, A.J. Reduction of simulated lunar glass by carbon and hydrogen and its implications for lunar base oxygen production. *Lunar Planet. Sci.*, **XXII**, 881–882 (Abstract) (1991).

[9] Meek, T.T.; Fayerweather, L.A.; Godbole, M.J.; Vaniman, T.; Honnell, R. Sintering lunar simulants using 2.45 GHz radiation. In: S.W. Johnson and J.P. Wetzel (eds.), *Engineering, Construction, and Operations in Space: Proc. Space '88*, pp. 102–110. American Society of Civil Engineers, Reston, VA (1988).

[10] O'Donnell, P.M. *Reactivity of Simulated Lunar Material with Fluorine*, NASA-TM-X-2533. NASA, Washington, D.C. (1972).

[11] Rosenberg, S.D. A lunar-based propulsion system. In: W.W. Mendell (ed.), *Lunar Bases and Space Activities of the 21st Century*, pp. 169–176. Lunar and Planetary Institute, Houston, TX (1985).

[12] Schmitt, H.H. *Return to the Moon: Exploration, Enterprise, and Energy in the Human Settlement of Space*, First Edition. Springer-Verlag, New York (October 2007).

[13] Simonds, C.H. Sintering and hot pressing of Fra mauro composition glass and the formation of lunar breccias. *American J. Sci.*, **273**, 428–439 (1973).

[14] Slyuta, E.N.; Abdrakhimov, A.M.; Galimov, E.M.; Vernadsky, V.I. The estimation of helium-3 probable reserves in lunar regolith. *Lunar and Planetary Science*, **XXXVIII** (2007).

[15] Steurer, W.H.; Nerad, B.A. Vapor phase reduction. In: W.F. Carroll (ed.), *Research on the Use of Space Resources*, NASA JPL Publ. 83-86. NASA Jet Propulsion Laboratory, Pasadena, CA (1983).

[16] Sviatoslavsky, I.N.; Cameron, E.N. *Geology of Mare Tranquillitatis and Its Significance for the Mining of Helium*, WCSAR-TR-AR3-9006-1, pp. 28–50. Wisconsin Center for Space Automation and Robotics (1990); Cameron, E.N., *Geology of Mare Tranquillitatis and Its Significance for the Mining of Helium*, WCSAR-TR-AR3-9006-1, p. 62. Wisconsin Center for Space Automation and Robotics (1990).

[17] Sviatoslavsky, I.N. *The Challenge of Mining He-3 on the Lunar Surface: How All the Parts Fit Together*, WCSAR-TR-AR3-9311-2, p. 12. Wisconsin Center for Space Automation and Robotics (1993).

[18] Taylor, L.A.; Oder, R.R. Magnetic beneficiation and hi-Ti mare soils: Rock, mineral, and glassy components. In: S.W. Johnson and J.P. Wetzel (eds.), *Engineering, Construction, and Operations in Space: Proc. Space '90*, pp. 143–152. American Society of Civil Engineers, Reston, VA (1990).

[19] Williams, R.J.; McKay, D.S.; Giles, D.; Bunch, T.E. Mining and beneficiation of lunar ores. *Space Resources and Space Settlements*, NASA-SP-428, pp. 275–288. NASA, Washington, D.C. (1979).

[20] Zhao, Y.; Shadman, F. Kinetics and mechanism of ilmenite reduction with carbon monoxide. *AlChE J.*, **36**, 443 (1990).

10

Lunar space tourism

It's been far too long. It's time to go back.
Eric Anderson, CEO, Space Adventures Ltd.,
announcing the availability of two commercial tickets to the Moon

In 2010 Virgin Galactic will commence suborbital operations, setting the stage for commercial manned spaceflight (Figure 10.1). Although flights to the Moon represent a whole different order of magnitude than suborbital, or even orbital flight, the flights of Virgin Galactic will represent a new impetus for space tourism. In the 21st century, thousands of wealthy tourists pay tens of thousands of dollars to climb Mount Everest, dive cave systems, and ski across the Arctic. Tourism is a proven market and one that will inevitably result in trips to the Moon, as evidenced by the fact that it is already possible to buy a ticket!

SPACE ADVENTURES LTD.

On August 10th, 2005, Arlington, Virginia–based Space Adventures Ltd., the world's leading space experience company, announced they would be selling $100 million tickets to the far side of the Moon. The company, which has organized spaceflights to the International Space Station (ISS) for spaceflight participants Dennis Tito, Mark Shuttleworth, Greg Olsen, and Anousheh Ansari, also offers a variety of space experience programs ranging from parabolic flights to cosmonaut training. Adding to its status within the space industry is the list of advisory members which includes Space Shuttle astronauts Kathy Thornton, Robert (Hoot) Gibson, Norm Thagard, Byron Lichtenberg, and Apollo 11 moonwalker Buzz Aldrin. The company described the mission, designated Deep Space Exploration-Alpha (DSE-Alpha), during a press conference.

Figure 10.1. Virgin Galactic's SpaceShipOne. Image courtesy: Scaled Composites.

Deep Space Exploration-Alpha

The DSE-Alpha mission will be the first of several lunar missions Space Adventures plans to sell to wealthy private space explorers over the next few years. The mission is possible thanks to Space Adventures' partnership with the Federal Space Agency of the Russian Federation and the Rocket and Space Corporation Energia (RSC Energia). Based on their research, Space Adventures has identified more than a thousand people worldwide who have the financial resources to pay for the most expensive holiday ticket ever, among them veteran spaceflight participant Greg Olsen.

The feasibility of such a mission has been evaluated by RSC Energia, which proposed two options. The first option requires a direct rendezvous and docking in low Earth orbit (LEO) with an upper-stage booster, while the second option involves a multi-day visit to the ISS followed by an upper-stage docking.

Since the Soyuz (Figure 10.2), designed in the 1960s, was originally intended for lunar missions, the modifications needed to re-certify it for lunar flights are well understood and may be implemented relatively easily. In the mission envisioned by Space Adventures Ltd., two private lunar explorers will fly to the Moon in a Soyuz spacecraft piloted by a Russian cosmonaut. Offering a spartan ten cubic meters of habitable volume the occupants of the Soyuz spacecraft will have no more living space than if they were inside an average SUV. While this may seem cramped, by spacecraft standards it is in fact quite spacious, especially when compared with the six

Figure 10.2. Soyuz spacecraft performing docking maneuver. Image courtesy: NASA.

meters of habitable volume the three-man crew of Apollo 8 had to live in during their six-day lunar mission in December 1968.

Apollo 8, as many are aware, was the first mission to fly around the Moon, and therefore shares a similar mission profile with DSE-Alpha. Many may argue paying $100 million for a fly-by trip to the Moon is a little steep, but the crew will enjoy spectacular views of the lunar landscape as the mission will be planned when the far side of the Moon is illuminated.

The DSE spacecraft and the missions

Direct Staged Mission

DSE-Alpha's baseline Direct Staged Mission (DSM) will commence with the launch of a Soyuz TMA spacecraft by a Soyuz 11A511U booster into LEO. The following day the Block DM Upper Stage (US) will be launched by a Zenit 3 LV into a parking orbit, after which the Soyuz TMA will begin to maneuver to rendezvous with the US. Day 3 will feature the Soyuz docking with the Block DM, after which the Block DM will fire its engines to send the Soyuz on a translunar trajectory. Following burn-out, the Soyuz will undock from the US and continue on its course toward the Moon. Days 4 and 5 will be spent traveling to the Moon with arrival scheduled for Day 6. Since the Soyuz will not have sufficient propellant to enter lunar orbit the mission will be a fly-by, affording the crew just a few hours to gaze out of the windows and take pictures. Two more days will be required for the return journey to Earth with re-entry scheduled for Day 9.

International Space Station Staged Profile

The ISS Staged Profile commences with a two-day journey to the ISS onboard a Soyuz. On Day 2 the Soyuz will dock with the ISS for a scheduled 10 to 14-day stay affording the crew an opportunity to tour the facilities and engage in a little sightseeing from the Cupola.

Towards the end of the ISS excursion a Zenit booster will be readied to deliver a high-energy US, designated the Block DM, basically an engine used to send the Soyuz spacecraft to the Moon. Once the Block DM is parked in LEO the crew will undock from the ISS and rendezvous with the US. On Day 15 the Block DM engine will fire, accelerating the Soyuz to 38,000 kmh. Once the fuel is spent, latches holding the spacecraft together will release, and the Soyuz will back away from the US. Following a two-day coast the Soyuz will enter lunar gravity for a fly-by of the Moon's far side, followed by another two-day coast back to Earth for re-entry on Day 21.

THE ARTEMIS PROJECT

The Artemis Project is a proposal by Artemis Society International (ASI) and the Lunar Resources Company (LRC) to establish a permanent, self-supporting manned lunar base. In Greek mythology Artemis, the Moon, is the twin sister of Apollo, the Sun. In addition to being the Goddess of the Moon, Artemis was also goddess of the hunt, an appropriate role given the expeditionary nature of the first flights to the Moon, requiring astronauts to live off the land. Intended to be a 100% privately funded commercial venture, the Artemis Project aims to develop lunar resources for profit and to demonstrate that manned spaceflight is within the reach of private enterprise.

Artemis reference mission

Although the Artemis Project does not specify which Launch Vehicle (LV) it intends to use, the reference mission suggests at least two launches will be required to deliver all the mission components to LEO.

In the proposed Artemis reference mission the Lunar Transfer Vehicle (LTV) will be attached to a habitat. The LTV, which will support the crew during the flight between Earth and lunar orbit, will follow a trajectory similar to the Apollo missions. Once the configuration arrives in lunar orbit, the habitat will separate from the LTV and land on the lunar surface, while the LTV remains, unoccupied, in lunar orbit.

During a seven-day stay on the lunar surface the crew will establish a lunar habitat, gather samples, and set up power systems and communication equipment. Another activity the Artemis astronauts will be involved in will be the filming of various aspects of the flight and activities performed during the surface stay. The footage will then be used to promote the idea of traveling to the Moon.

The two-hour ascent to lunar orbit is performed in an open vehicle, which means Artemis astronauts will be completely reliant on their space suits for life support. Once the ascent stage has docked with the LTV, rockets are fired to place the configuration on a trajectory to Earth orbit. Once in LEO the LTV is parked on the ISS or remains in a parking orbit, either to be used on a later mission or to be leased to private companies. Once the LTV has been secured, the crew return to Earth in whatever vehicle brought them to orbit.

Although the Artemis Project does not specify how their astronauts will reach and return from LEO it is likely they may take advantage of a crew-rated Delta IV, a Soyuz, or perhaps one of the LVs offered by private space companies such as Blue Origin. Whereas the reference mission is a little short on detail regarding the crew vehicle to and from orbit the Artemis Project has a reasonably well-developed idea of how the three spacecraft achieve the goal of ferrying crew to the Moon.

Artemis mission architecture

Lunar Transfer Vehicle

The LTV will be designed to function as an orbiting habitat capable of supporting the crew between the Earth and the Moon. Based on the design of the Spacehab module built by McDonnell Douglas, the LTV will feature a propulsion package positioned at the opposite end to the hatch. To enable a trajectory capable of sending the LTV from LEO to a lunar orbit the propulsion will need to impart a delta-V of 4,119 m/s to the Artemis stack. Inside the module will be the standard systems for electrical power, thermal control, and life support sufficient to support a crew for at least the three-day outbound and three-day inbound trip. External features will include radiators, antennae, a Reaction Control System (RCS), and sensors to support the communication suite. Refueling of the LTV will be achieved simply by exchanging fuel tanks, launched while the LTV loiters in LEO. Other components of the LTV will include a Forward Service Module (FSM), Aft Service Module (ASM), and Propulsion Module, which may be launched individually and assembled on orbit or sent to LEO as an integrated package.

Descent Stage

The Descent Stage, consisting of six cylindrical hydrogen tanks surrounding one spherical oxygen tank, will serve as a propulsion module used only once for landing on the Moon, after which the fuel tanks will be used for storing oxygen mined from the lunar regolith. The tank structure, surrounded by a hexagonal cage structure, will be sized to fit inside the payload shroud of a Titan IV LV. In the launch configuration the tanks would probably not be visible as they would be covered with gold foil to reflect sunlight away from the cryogenic systems.

Lunar Exploration Base

The Lunar Exploration Base (LEB) comprises the manned element similar to the Spacehab modules flown on the Space Shuttle. The Descent Stage and LEB config-uration is intended to land vertically on the Moon, after which the LEB is rotated to lie horizontally to provide a more astronaut-friendly layout. The term astronaut-friendly is perhaps not the first descriptive term that comes to mind when describing the Ascent Stage, given this vehicle is an open vehicle with astronauts wearing just their extravehicular activity (EVA) suits for protection during the two-hour ascent to orbit and rendezvous phase. Although to some observers this may seem a little risky, the Artemis Project Team argue astronauts should not be exposed to any greater danger than during a two-hour excursion on the lunar surface. The Ascent Stage will be required to impart a delta-V of 1,869 m/s to ensure the vehicle, its crew, and lunar samples attain a 100 km lunar orbit.

Propellant

A quandary the Artemis Project Team faced was deciding which fuel would be the safest and still be capable of providing the necessary I_{SP}. Although low-efficiency hypergolic fuels are relatively safe, reliable, and easy to store, the I_{SP} yield is relatively low compared with cryogenic fuels such as liquid hydrogen (LH_2) or liquid oxygen (LOX). LH_2 and LOX, although yielding a higher I_{SP}, are prone to boil-off, requiring more components and operations for successful engine starts, and are far less reliable than hypergolic fuels. In the end the Artemis Project Team decided to selected the low I_{SP} yield and reliability of hypergolic fuels, choosing unsymmetrical dimethyl-hydrazine (UDMH) and N_2O_4 (nitrogen tetroxide).

The stack

The LTV, LEB, and Descent Stage comprise the stack of three spacecraft that will first be assembled in LEO or at the ISS before flying to the Moon on a trajectory similar to the one used by the Apollo spacecraft.

Artemis crew selection

The goal of the Artemis Project is to sell the idea of flying to the Moon. Given this aim, the crew selection criteria will be rather different from the conventional astro-naut select-in standards. Although fitness, medical, and technical skill requirements similar to those required by orbital spaceflight participants and professional astronauts will probably apply, Artemis crewmembers will also be required to possess public-speaking, photography and even acting experience. These skills will enable Artemis astronauts to bring the experience of being on the Moon home to those on Earth.

BIGELOW AEROSPACE

Perhaps the first general contractor for the final frontier will be Las Vegas billionaire Robert T. Bigelow's company, Bigelow Aerospace. The first step towards that goal came in July 2006 when a Russian booster launched Bigelow's Genesis I prototype module into orbit (Figure 10.3). The next module, Genesis II, was successfully launched a year later, and it will be followed by Sundancer, a human-rated module, in 2010. Sundancer will be followed by the largest of Bigelow's modules, the BA330, at which point Bigelow hopes to start charging $5 million a ticket for a week-long stay in LEO. For Bigelow however, sending wealthy spaceflight participants into LEO merely constitutes "Job 1" and serves only as a stepping stone to first establishing a construction zone at Lagrange Point 1 (L1), where habitats would be built and then landed on the Moon.

Ironically, NASA, who pioneered the inflatable space habitat technology (Bigelow licensed the technology, now known as the Transhab System), is now discussing its options with Bigelow regarding building lunar infrastructure. Based on his extensive experience as a contractor, Bigelow may be the ideal person to take on the task of building structures on the Moon, as evidenced by some of the remarks he made in an exclusive interview with Alan Boyle in February, 2007.

"If people have ever been around a construction site at night, they'll see a bunch of lights on those machines, and some service trucks there. Those service trucks

Figure 10.3. Bigelow Aerospace's BA330 Module. Image courtesy: Bigelow Aerospace.

aren't there just because there's nothing better to do than visit the machinery. It's because that machinery breaks down constantly on Earth, all of the time. Every construction site has that feature to it. People who have never been to construction sites are completely unaware that this is a habitual problem on Earth, let alone the Moon."

The problem Bigelow was alluding to was on the Moon, as a result of the problematic lunar dust the potential for machinery failure will be exacerbated, a situation which can be resolved by having no machinery since Bigelow's habitats will be built ready to be deployed at his L1 construction sites.

"It has been said, 'If God wanted man to fly, He would have given man wings.' Today we can say, 'If God wanted man to become a spacefaring species, He would have given man a moon."

Krafft Ehricke, German American space visionary

In the long term, the Moon, like the summit of Mount Everest, will become just another destination only a space cellphone call away. Although the possibility of the lunar landscape becoming a tempting travel destination for tourists cannot be realized until the necessary facilities are constructed, the activities of Space Adventures and the first forays of Constellation point strongly to the prospect of lunar tourism becoming as viable as adventure travel is today.

Epilogue

Now we have the capability to leave the planet, and I think we should give careful consideration to taking that option. Man has always gone where he has been able to go, it is a basic satisfaction of his inquisitive nature, and I think we all lose a little bit if we chose to turn our backs on further exploration. Exploration produces a mood in people, a widening of interest, a stimulation of the thought processes, and I hate to see it whither. Our Universe should be explored by microscope and by telescope, but I don't believe the argument that less emphasis on one will cause a more powerful focus on the other. When man fails to push himself to the possible limits of the Universe in a physical sense, I think it causes a mental slackening as well, and we are all the poorer for it. Space is the only physical frontier we have left, and I believe its continued exploration will produce real, if unpredictable benefits to all of us who remain behind on this planet. That one cannot spell out in any detail what these benefits will be does not contradict or deny their existence.

Michael Collins, Apollo 11 astronaut [1]

Almost 15 years after the Senior President Bush's announcement of the Space Exploration Initiative (SEI) his son declared his administration also had an ambitious space plan. The Vision for Space Exploration (VSE) that intends to return astronauts to the Moon and then journey on to Mars may echo the goals of the SEI, but there are important differences suggesting the VSE will be successful.

Whereas the SEI paid lip service to the concept of utilizing lunar *in situ* resources the VSE places this role directly into the critical path of the development of the lunar outpost.

"The fundamental goal of this vision is to advance U.S. scientific, security, and economic interests through a robust space exploration program."

This statement subordinating space exploration to the primary goals of scientific, security, and economic interests identifies the benefits against which the costs of exploration can be weighed. Not only is this important for policymaking due to the sharing of science, security, and economic dimensions by federally funded activities, it also appeals to the general public who are more inclined to spend tax dollars on an economic enterprise than on a plan simply intended to send astronauts to the Moon and return them safely. Similarly, the economic and technical linkages between the effort of establishing the lunar outpost and the greater good on Earth represent a powerful argument for pursuing the goal of space development.

In addition to identifying the economic potential of returning to the Moon, the VSE's intention of establishing a permanent lunar outpost suggests the potential for such an installation to build a self-sustaining off-planet foothold. By living on the Moon it will be possible to learn how to generate oxygen from rocks and process *in situ* materials that may be crucial to lowering the cost of traveling to Mars. Furthermore, technologies developed on the Moon such as the closed-loop or nearly closed–loop industrial infrastructure not only serve to emplace materiel required to make our civilization a space-faring one but may ultimately serve as a model for the much larger one here on Earth. Someone who committed his life to the exploration and development of space was legendary NASA Flight Director Gene Kranz, who offered this missive to the Constellation Program workforce at Johnson Space Center in April 2008:

"What you're looking at here is America's drive to explore the Moon, a space program coming back to life, a re-awakening, a renaissance, a re-dedication of our nation that will lead the way into the next chapter of human destiny.

It's very exciting stuff, but you already know that, because you are the folks that are making it happen. I salute all of you that are on the team that will take us back to the Moon and on to Mars.

You are today where I was about forty-five years ago, when we decided to go to the Moon for the first time, when we responded to President Kennedy's challenge 'we choose to go the Moon in this decade and do the other things, not because they are easy, but because they are hard.' You have a similar challenge, given to us by President Bush, and it is up to you to make it happen.

There are many similarities between our challenge to reach the Moon back in 1961 and your challenge today. The real similarity is the human factor, the people, the learning curve that you have in place that started many years earlier in a place called Cape Canaveral.

We have a marvellous array of technology, and text books are written. You have flight experience, and you have all the tools necessary to make this happen.

When we started in 1960, we had to write the books, we had to invent the technology. They are there, they are available to you, and you can make this next program happen.

Our learning curve, as we went through Mercury, Gemini and Apollo was very steep. With Mercury we learned of leaders and leadership—leaders must have integrity, they are up-front, they challenge the team to accomplish their mission.

We also learned a lot about ourselves with Mercury, as a lot of us came in from aircraft testing, with egos as big as the room we worked in. We had to learn to leave our egos at the door and become a team, so we became one.

With Gemini, we had to cope with the new technologies of space, with fuel cells, computers and bi-propellant rocket engines. But again, after we had met the challenge of these new technologies, we continued to learn.

We learned discipline, a focus upon our objective that was so intense that we would never fail to achieve that objective.

We also learned the value of high morale, because with Mercury, we knew we were now succeeding because of our beliefs, in our mission, our team and ourselves. So as we entered Apollo we had all the tools in place, as a team, to reach for the Moon.

But we were bloodied by the Apollo I fire, and I hope you should never have to go through a day like we did.

There we became tough and competent. Tough meant we were never going to walk away from our responsibilities, because we're forever accountable for what we do, or what we fail to do. Competent, because we're never going to take anything for granted, we're never going to stop learning. From now on, the teams of Apollo are going to be perfect.

The challenges you will face are going to be enormous, but you have history to provide you with the direction you need. You have leaders in place today, within this agency who have come through the program. They have faced these challenges before, they know how to respond to them.

You have an American public that is looking for you to succeed. They believe in you, they believe in space—and it is up to you to make these beliefs come true.

You have the Moon and Mars that you can reach for. It's a star, it's a mystery, but if you set your aims high, you can accomplish any challenge you are faced with. But most of all you must believe in yourself, and you must believe in your team.

If you have that belief, you can make it happen."

The decision the United States made more than 35 years ago to retreat from the Moon and Mars was a mistake which cost a generation its chance to make a difference. For far too long humans have been restricted to low Earth orbit (LEO) operations and now it really is time to move beyond LEO and start some real exploration. Cynics may complain the Constellation Program is too expensive, but the reality is the average American is spending only *15 cents per day* on the space program. Returning to the Moon will allow NASA to restore its core values and once again enable the agency to pursue goals that raise the human spirit and reach for the stars once again. Furthermore, the Constellation Program represents an opportunity for the United States and NASA to reawaken the spirit of adventure, inspire innovation, and reenergize space exploration as it will surely do when astronauts once again set foot on the Moon.

REFERENCE

[1]   Collins, M. *Carrying the Fire: An Astronaut's Journey*. Farrar, Strauss, & Giroux, New York (1974).

Glossary

Combustion chamber A *combustion chamber*, a *nozzle*, and an *injector* are the main elements of a typical rocket motor. The burning of propellants occurs inside the combustion chamber, which must be extremely strong to contain the tremendous pressures and extraordinarily high temperatures (~3,300°C). The chemical–thermal energy generated in the combustion chamber is converted into kinetic energy by the nozzle, which converts the high-pressure, high-temperature gas in the combustion chamber into high-velocity gas of lower pressure and temperature.

Composite propellants This propellant category utilizes an identification system based on the type of polymeric binder used. Common binders used in the design of rockets include polybutadiene acrylic acid acrylonitrile (PBAN) and hydroxyterminator polybutadiene (HTPB). The most common fuel is aluminum, and ammonium per-chlorate is usually used as an oxidizer, a combination used in the solid rocket boosters (SRBs) that are strapped onto the Space Shuttle.

Cryogens Cryogenic propellants, or cryogens, are liquefied gases such as liquid hydrogen (LH_2) and liquid oxygen (LOX) stored at extremely low temperatures. For example, LH_2 is in a liquid state at −253°C and LOX is liquid at −183°C. Not surprisingly, because of these super low temperatures, LH_2 and LOX are difficult to store, a problem compounded by LH_2's low density (0.071 g/mL), requiring a large storage volume, although this drawback is compensated by LH_2 delivering an I_{SP} more than 30% more than most other rocket propellants. An example of an LV using liquid propellants is the Space Shuttle, which uses LH_2 as the fuel and LOX as the oxidizer.

Drag Although drag is encountered mostly during launch and re-entry, an orbiting spacecraft will still experience drag as it moves through the Earth's upper atmosphere. In fact, if a spacecraft were to descend to an altitude of 160 km, atmospheric drag would bring it down in a matter of days. This deterioration of a spacecraft's orbit due to drag is referred to as *decay* and must be compensated by firing of rockets at various intervals to boost an orbiting spacecraft's altitude, a process known as an *orbital maneuver*.

Earth orbit rendezvous   Earth Orbit Rendezvous (EOR) is a mission architecture component requiring a docking of two spacecraft in LEO. For example, the Constellation mission architecture will require unmanned mission elements to be launched on a cargo vehicle and the manned elements to be launched on the crew LV. In LEO the unmanned and manned elements will rendezvous and dock before proceeding to the Moon.

Escape velocity   For a spacecraft to reach the Moon it must attain an escape velocity of 11.2 kilometres per second, which is the velocity required to escape Earth's orbit. Normally, for an actual escape orbit, a spacecraft is first placed into LEO and then accelerated at that altitude.

Human-rating requirements   The process of human-rating a launch vehicle is governed by NASA's Office of Safety and Mission Assurance publication NPR 8705.2. NPR 8705.2 is also known as the *NASA Human Rating Requirements Compliance Verification Guide*, a technical document designed to provide the maximum reasonable assurance of safety for people involved with spaceflight activities.

NPR 8705.2 criteria

A human-rating plan must include:	*The Human Rating Board consists of:*
Design criteria	Associate Administrator for Space Operations
System designs	Associate Administrator for Space Explorations
Test requirements and procedures	Chief Safety and Mission Assurance Officer, Office of Primary Responsibility
Software design	
Test and verification requirements	Chief Medical and Health Officer, Independent Technical Authority
System safety and reliability engineering	Chief Engineer, Independent Technical Authority
Human factors engineering requirements	
Health requirements	

NPR 8705.2. *NASA Human Rating Requirements: Compliance Verification Guide.* NASA Office of Safety and Mission Assurance, Washington, D.C. (August 29, 2005).

NPR 8705.2 covers the spectrum of capabilities and requirements pertaining to human-rating an LV and serves as a rigorous safety mechanism ensuring spaceflight systems are certified in compliance with acceptable risk management parameters.

Hypergolic The distinguishing property of this fuel category is the spontaneous ignition of the fuel and oxidizer on contact with each other. This means a system utilizing hypergols can easily be started and re-started, making hypergolic propellant ideal for spacecraft maneuvrring systems. Another bonus for rocket engineers is that hypergols remain liquid at normal temperatures so there are no storage problems, although the high toxicity of this type of propellant means it must be handled carefully.

Examples of hypergolic fuels include hydrazine and monomethyl hydrazine (MMH), and unsymmetrical dimethyl hydrazine (UDMH). Although hydrazine provides superior performance as a rocket fuel it cannot be used as a coolant due to its high freezing point, whereas MMH has a lower freezing point while still providing good performance as a fuel.

Inclination, periapsis, and apoapsis An orbiting spacecraft will be oriented at a certain *inclination* relative to the Earth. For example, an inclination of $0°$ (also termed a low-inclination orbit) indicates the spacecraft is in orbit about the Earth's equator whereas a spacecraft that has an inclination of $90°$ (this would be termed a high-inclination orbit) indicates it is in orbit about the Earth's poles. Regardless of the type of inclination, an orbiting spacecraft will follow an oval-shaped path known as an *ellipse*, which means there will be a point of closest approach to Earth, known as *periapsis*, and a point farthest away from Earth, known as *apoapsis*.

Interplanetary trajectory and transfer orbits To send a spacecraft to the Moon, it isn't sufficient to achieve escape velocity and simply point at the Moon. The spacecraft must be launched on a trajectory/transfer orbit that will intersect the Moon. Due to the requirement of considering the gravitational interactions of not only the Earth and the Moon, but the Sun as well, the complexities associated with calculating transfer orbits appear computationally overwhelming.

Liquid propellants A liquid propellant system is characterized by storing the fuel and oxidizer in separate tanks from which a series of valves, pipes, and turbopumps feed the fuel and oxidizer to the combustion chamber where they burn to produce thrust. One advantage of a liquid propellant system is its throttleable capability, achieved by controlling the propellant flow to the combustion chamber. This design characteristic also permits stopping and re-starting the engine.

Low Earth orbit To attain Earth orbit, a spacecraft must be launched to an altitude above the Earth's atmosphere and accelerated to orbital velocity, which is between 6.9 km/s and 7.8 km/s. The inclination of the orbit desired will have a bearing upon the amount of propellant required, with low-inclination orbits being the most energy-efficient and high-inclination orbits demanding the most propellant.

To reach a low-inclination/energy-efficient orbit a spacecraft is launched in an eastward direction from a site as close as possible to the equator. Such a direction and location confers an advantage on the launching spacecraft thanks to the contribution the rotational speed of the Earth makes to the final orbital speed of the spacecraft.

This is one reason NASA's primary launch site, Cape Canaveral, Florida, is located where it is, since the relatively low latitude of 28.5°N results in a free ride equating to adding 1,471 km/h to any eastward-launched orbital spacecraft.

Due to the rapid orbital decay of objects in orbits below 200 km, the accepted definition of low Earth orbit (LEO) is between 160 km and 2,000 km above the Earth's surface. For example, the International Space Station (ISS) orbits the Earth in an LEO varying between 319.6 km and 346.9 km altitude.

Lunar orbit insertion   Lunar orbit insertion (LOI) describes the process of a spacecraft slowing down as it approaches the Moon. As a spacecraft approaches the Moon lunar gravity begins to supercede Earth's gravity, and to achieve orbital velocity around the Moon the vehicle must retrofire (rockets facing in the direction of motion) its rockets.

Lunar orbit rendezvous   Lunar orbit rendezvous (LOR) is a mission architecture component requiring a docking of two spacecraft in lunar orbit. In the Constellation architecture the Crew Vehicle will be parked unmanned in lunar orbit while the crew descend to the lunar surface in the Lunar Surface Access Module (LSAM). After seven days on the lunar surface the LSAM will return the crew and rendezvous and dock with the Crew Vehicle, whereupon the crew will transfer back into the Crew Vehicle for the return to Earth.

Nozzle   The aim of rocket engineers when designing the engine is ensuring the nozzle is long enough to permit the reduction of pressure in the combustion chamber at the nozzle exit to the pressure existing outside the nozzle. This is important since thrust is at the maximum when the pressure at the nozzle exit is equal to the ambient outside pressure, a design feature referred to as *optimum expansion*. Another factor must be considered as the rocket ascends to orbit since the discharge of exhaust gases is limited by the separation of the jet from the nozzle wall. Although this process occurs normally when the rocket is close to sea level pressure, once the rocket reaches lower pressures the separation does not occur and there is a requirement for a different type of engine and nozzle.

Orbit perturbations   The launch of a spacecraft comprises a period of powered flight commencing with lift-off from the launch pad and concluding upon burnout of the rocket's last stage when the vehicle is in orbit. Once in orbit the spacecraft is considered to be in a state of free flight and its trajectory is subject only to the gravitational pull of the Earth. If the spacecraft travels far from the Earth, its trajectory may be affected by the gravitational influence of the Moon or other planets, an effect referred to as *third-body orbit perturbations*. Other orbit perturbations affecting the spacecraft include *drag*.

Orbital maneuvers   In addition to permitting spacecraft to perform altitude changes, orbital maneuvers are used to transfer from one type of orbit to another, to permit rendezvous with other spacecraft, and to change the parameters of an exiting orbit to meet a launch window. Perhaps the most complicated and most precise orbital maneuver is orbital rendezvous, which requires a spacecraft to intercept another spacecraft or object at a rendezvous point at the same time.

Orbital mechanics   The motions of satellites and spacecraft moving under the influence of gravity, atmospheric drag, and thrust, are described by orbital mechanics, the root of which lies in Newton's law of universal gravitation. Typical applications of orbital mechanics include the calculation of ascent trajectories, re-entry and landing, orbital rendezvous, and interplanetary trajectories.

Parallel staging   This type of staging is familiar to anyone who has watched a Space Shuttle launch. In parallel staging several small first stages are strapped onto a central core booster or a central sustainer rocket. At launch, all the engines are ignited. When the propellant in the strap-on stages is exhausted, the strap-ons are discarded and the central sustainer rocket continues to burn, carrying the payload into orbit. Unlike serial staging, the discarded rocket boosters are retrieved and re-used.

Rocket propulsion   The principle of rocket operation is described by Newton's third law of motion that states "for every action there is an equal and opposite reaction." Propellants are delivered to the combustion chamber where they react chemically to generate hot gases which are accelerated and ejected at high velocity through a nozzle. The forceful expulsion of hot gases through the nozzle provides momentum or the thrust force of the rocket motor in a similar manner to a gun recoiling after being fired.

Serial staging   Serial staging was used on the three-stage Saturn V system that took astronauts to the Moon. Normally, in this type of staging a small, second-stage rocket is placed on top of a large first stage (FS). When the propellants of the FS are exhausted its engines are extinguished and the FS and second stage separate. The FS is discarded and usually not recovered and the second-stage engine ignites and carries the payload to orbit.

Space architecture   When mission planners discuss space architecture they are referring to the three components comprising a mission design. These components are the space segment, the launch segment, and the ground segment.

   The *space segment* includes a description of the orbital mechanics required to place a spacecraft into a particular orbit such as low Earth orbit (LEO) and the means of planning planetary surface systems such as an outpost on the Moon. It also describes concept design, analysis, planning, and integration of systems hardware for spacecraft in addition to explaining spacecraft/habitat layout, crew quarter design, cabin architectures and human factors analysis. This segment also describes orbital maneuvering, orbital perturbations, and orbital maintenance.

   The *launch segment* describes the launch profile of the LV from the time it leaves the launch pad to the time it enters space. It also describes vehicle-to-ground communications interface design, mission parameters, and mission geometry. The launch segment also explains the process of selecting launch windows.

The *ground segment* includes all the elements required to put the spacecraft onto the launch pad and includes the process of selecting launch systems, processing facilities, and operational usage assessments. Also included in this segment is an explanation of the constraints of the design process, spacecraft configuration considerations, design budgets, and integration of the spacecraft design. The ground segment may also describe factors such as mission operations plans, mission operations functions, and qualification programs.

Specific impulse   In addition to thrust, rocket engineers are also interested in delivering the highest engine performance, a measure known as the *specific impulse* (I_{SP}) of a rocket. I_{SP} is measured by how many kilograms of thrust is provided by burning one kilogram of propellant in one second. Sometimes this value is difficult to calculate for different phases of a mission because the I_{SP} has a different value on the ground than in space because the ambient pressure is a factor affecting thrust, which in turn is a factor in calculating I_{SP}. Compounding the problem of accurately calculating I_{SP} are the losses occurring within the rocket engine such as the nozzle and pumps, a loss rocket engineers attempt to mitigate by designing efficient engines and nozzles.

Staging   Most of the weight of a Launch Vehicle is propellant. As propellants are burned during the powered ascent phase of the mission, a larger proportion of the vehicle's weight becomes empty. Once the propellant tanks are empty it makes sense to discard them to lighten the weight of the LV thereby helping it achieve orbital velocity. The process of discarding propellant tanks is called staging.

Technology Readiness Levels   To assess the maturity of evolving technologies required to develop a new LV, NASA utilizes Technology Readiness Levels (TRLs). Some of the technologies conceptualized for the LVs and spacecraft described in this book were not suitable for immediate application. To certify the new technologies, NASA had to conduct experimentation, refinement, and realistic testing until the technologies were sufficiently proven. This process followed a nine-step plan as outlined in the table at the top of the page opposite.

Thrust   The propulsive force that sends the rocket skywards is termed *thrust* and is generally measured in either kilograms or Newtons. A typical rocket engine configuration features a combustion chamber with a nozzle through which the gas is expelled. The key feature to note is that pressure distribution within the chamber is not symmetrical. Inside the chamber the pressure variation is very little, but closer to the nozzle the pressure decreases. The force generated by the rocket is a function of the force due to gas pressure on the bottom of the chamber not being compensated from outside the chamber, thereby creating a pressure differential between the pressure internally and externally. This pressure differential, or thrust, is opposite to the direction of the gas jet and therefore pushes the chamber upwards.

NASA Technology Readiness Levels.

Technology Readiness Level	Description
1. Basic principles observed and reported	Lowest level of technology maturation; scientific research begins to be translated into applied research
2. Technology concept formulated	Practical applications of technology identified; no experimental proof to support concept at this stage.
3. Analytical and experimental/ proof of concept	Research and development initiated; includes analytical studies to set technology into appropriate context; studies and experiments constitute "proof of concept"
4. Component validation in laboratory environment	Technological elements integrated; ensure all the elements work together; validation must support original concept and be consistent with requirements of potential system applications
5. Component validation in relevant environment	Fidelity of component increased significantly; technological elements integrated with realistic supporting elements, so total applications may be tested
6. System/subsystem prototype demonstration	Representative prototype system tested in a relevant environment; in many cases the only relevant environment is space
7. System prototype demonstration in space	Actual system prototype demonstration in space environment; prototype should be near or at scale of planned operational system
8. Actual system completed and "flight-qualified"	End of system development for most technology elements; may include integration of new technology into existing system
9. Actual system "flight-proven" through successful mission operations	Integration of new technology into existing system; does not include planned product improvement of ongoing or reusable systems

Mankins, J.C. *Technology Readiness Levels*, a White Paper. Advanced Concepts Office, Office of Space Access and Technology, NASA, Washington, D.C. (April 6, 1995).

Trans-Earth injection   Trans-Earth injection (TEI) describes the propulsive maneuver used to place a spacecraft on a trajectory that will intersect the Earth. Prior to the TEI burn the spacecraft is in a parking orbit around the Moon. The burn is timed to ensure the mid-point of the TEI is opposite the Earth.

Translunar injection   A translunar injection (TLI) is a propulsive maneuver used to place a spacecraft on a trajectory that will intersect the Moon. Prior to TLI burn the spacecraft is in a parking orbit around the Earth traveling at a velocity of approximately 28,150 kmh. Following the TLI burn, which is performed by a powerful rocket engine, the spacecraft will be traveling at approximately 39,400 km/h. The burn is timed very precisely to ensure the mid-point of the TLI is opposite the Moon.

Index